POLYMER SCIENCE A

INORGANIC POLYMERS CONTAINING -RE(CO)3L+ PENDANTS

Polymer Science and Technology

Additional books in this series can be found on Nova's website under the Series tab.

Additional E-books in this series can be found on Nova's website under the E-books tab.

POLYMER SCIENCE AND TECHNOLOGY

INORGANIC POLYMERS CONTAINING -RE(CO)3L+ PENDANTS

EZEQUIEL WOLCAN
AND
MARIO R. FÉLIZ

Novinka
Nova Science Publishers, Inc.
New York

LIBRARY OF CONGRESS CATALOGING-IN-PUBLICATION DATA

Available upon Request
ISBN: 978-1-61668-928-5

Published by Nova Science Publishers, Inc. ✝ New York

CONTENTS

PREFACE

We describe the synthesis, photophysical and photochemical properties of inorganic polymers containing $-Re(CO)_3L^+$ pendants with general formula $[(vpy)_2\text{-}vpyRe(CO)_3\ L^+]_{n\sim200}$ (vpy = vinylpyridine, L=α–diimine). Marked differences were found between the photochemical and photophysical properties of the polymers and those of the related monomeric complexes, $pyRe(CO)_3L^+$ (py = pyridine). The main cause of these differences is the photogeneration of the metal-to-ligand charge transfer (MLCT) excited sates in concentrations that are much larger when $-Re(CO)_3\ L^+$ chromophores are bound to $(vpy)_{n\sim600}$. This is the photophysical result of Re(I) chromophores being crowded in strands of a polymer instead of being homogeneously distributed through solutions of a $pyRe(CO)_3L^+$ complex. The photochemical and photophysical properties of polymers $\{[(vpy)_2\text{-}vpyRe(CO)_3L]CF_3SO_3\}_{n\sim200}$ and the related monomers $CF_3SO_3[pyRe(CO)_3L]$ (L = phen and bpy) were investigated in solution phase. The yield of formation and the kinetics of decay of the MLCT excited state were found to be dependent on medium and laser power. MLCT excited states in the polymers undergo a more efficient annihilation and/or secondary photolysis than in the monomers. Solvent and temperature effects on the polymers photophysical properties are rationalized in terms of the transition between rigid rod and coil structures of the Re(I)-polymers. Transmission electron microscopy (TEM) and dynamic light scattering (DLS) studies on acetonitrile solutions of the polymer $\{[(vpy)_2\text{-}vpyRe(CO)_3(bpy)]CF_3SO_3\}_{n\sim200}$ demonstrated that the Re(I) polymer molecules aggregate to form spherical nanodomains of radius $R \sim 156$ nm. Coordination of Cu(II) species to the Re(I) polymer causes a decrease in the nanodomain radius and a distortion from the spherical shape as well as a quenching of the MLCT excited state by energy transfer processes that are more efficient than those in the quenching of the monomer $CF_3SO_3[pyRe(CO)_3(bpy)]$ luminescence by Cu(II). Energy transfer between MLCT(Re→tmphen) and MLCT(Re→NO_2-phen) excited states inside mixed polymers like $\{[(vpy)_2\text{-}vpyRe(CO)_3(tmphen)^+]\}_n\{[(vpy)_2\text{-}vpyRe(CO)_3(NO_2\text{-}phen)^+]\}_m$ was evidenced by steady state and time-resolved spectroscopy. Current Förster resonance energy transfer theory was successfully applied to energy transfer processes in these polymers. The photochemical and photophysical properties of the polymers $[(vpy\text{-}CH_3^+)_2\text{-}vpyRe(CO)_3(phen)]_{n\sim200}$ have been investigated in solution phase and compared to those of a related polymer, $[(vpy)_2\text{-}vpyRe(CO)_3(phen)]_{n\sim200}$, and monomer, $pyRe(CO)_3(phen)^+$. Irradiations at 350 nm induce intrastrand charge separation in the peralkylated polymer, a process that stands in contrast with the energy migration observed with $[(vpy)_2\text{-}vpyRe(CO)_3(phen)]_{n\sim200}$. Electronically excited $\text{-}vpyRe(CO)_3(phen)^+$ chromophores and charge separated intermediates react with neutral species, e.g., 2,2′,2′′-nitrilotriethanol (TEOA), and anionic electron donors, e.g., SO_3^{2-} and I^-. The anionic electron donors react more efficiently with the MLCT excited state of these polyelectrolytes than with the excited MLCT state of $pyRe(CO)_3(phen)^+$.

Chapter 1

INTRODUCTION

The photochemistry of organometallic compounds represents an important branch of coordination chemistry. The interest in this research field is based on fundamental aspects as well as applications. Luminescent transition metal complexes have been utilized as photosensitizers in areas such us solar energy conversion, electron transfer studies, chemiluminescent and electroluminescent systems, binding dynamics of heterogeneous media and probes of macromolecular structure [1]. In this regard, the spectroscopy, photochemistry and photophysics of Re(I) carbonyl–diimine complexes *fac*-$Re(L)(CO)_3(\alpha\text{-diimine})^{0/+}$ (figure 1) continue to attract much research interest ever since their intriguing excited state properties were first recognised in the mid-1970s [2].

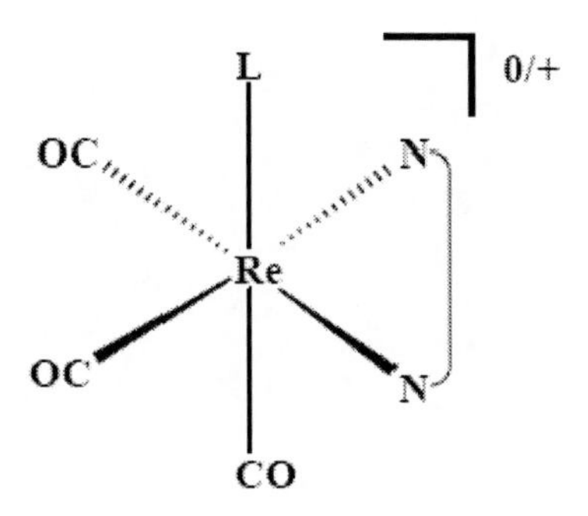

Figure 1. Structures of the $Re(L)(CO)_3(\alpha\text{-diimine})^{0/+}$ complexes. (The Re complexes are neutral or cationic for anionic or neutral axial ligands L, respectively).

Depending on the nature of the axial L ligand *fac*-Re(L)$(CO)_3$(α-diimine)$^{0/+}$ complexes are often strong luminophores, either in fluid solutions or in low-temperature glasses. The accessible excited states, Re(I) to azine metal-to-ligand charge transfer (MLCT), ligand-to-ligand charge transfer (LLCT), and/or intraligand (IL) excited states, are generally involved with the observed luminescence of these complexes at room temperature. A rational design in the synthesis of the α-diimine ligands was used to tune the photophysical and photochemical properties of the metal complexes in order to obtain photosensitizers that might be utilized in areas such as electron transfer studies [3] solar energy conversion [4–6] and potential technical applications to catalysis [7]. Possible applications as luminescent sensors [8–11] or molecular materials for non-linear optics [12,13] or optical switching [14] are also emerging.

Another approach involved the synthesis of compounds where the metal complex containing the chromophore unit was attached to an organic polymeric backbone. Much of the literature examples concerning inorganic polymers involve transition metal ions metalating poly(p-phenylenevinylene) polymers incorporating 2,2-bipyridines, polypyridil Ru(II) and/or Os(II) derivatized polystyrene [15] and some multimetallic oligomeric complexes containing Ru(II) and Os(II) coordinated to 1,10-phenanthroline [16].

However, up to year 2000, little attention was paid to Re(I) polymeric complexes. We have applied ligand substitution reactions of the Re(I) tricarbonyl complexes to the synthesis of a series of polymers derived from poly-4-(vinylpyridine) (figure 2, left).

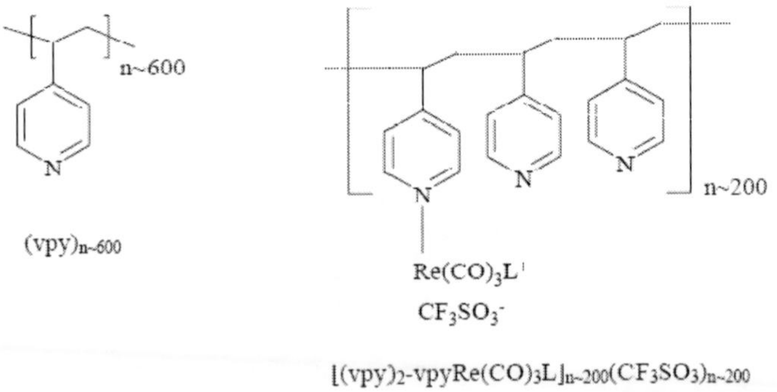

n~600

N

$(vpy)_{n\sim600}$

N N N

n~200

$Re(CO)_3L'$

$CF_3SO_3^-$

$[(vpy)_2\text{-}vpyRe(CO)_3L]_{n\sim200}(CF_3SO_3)_{n\sim200}$

Figure 2. Structural formulae of poly-4-(vinyppyridine) (left) and polymers derived from poly-4-(vinylpyridine) and $-Re(CO)_3(L)^+$ pendants (right) and the abbreviations used.

Poly-4-(vinylpyridine), represented hereafter as $(vpy)_{n\sim600}$, has a molecular weight $M_w\sim6{,}0x10^4$ and, on the average, it contains about 600 vinylpyridine (vpy) units per formula. These inorganic polymers are synthesized by attaching pendant chromophores $–Re(CO)_3(L)^+$ to one third of the pyridines of the organic backbone (figure 2, right). The polymers can be assigned the formal structure of figure 2 where pendant $-Re(CO)_3L^+$ groups are randomly distributed along the strand of polymer with an average of two 4-vinylpyridine groups for each one coordinated to a Re(I) chromophore. Molecular weights of the final polymers range from 1.7 to $2.0x10^5$ depending on the nature of the ligand L used in the synthesis of the polymers (figure 3).

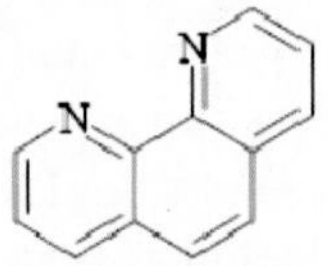

1,10 phenanthroline (phen)

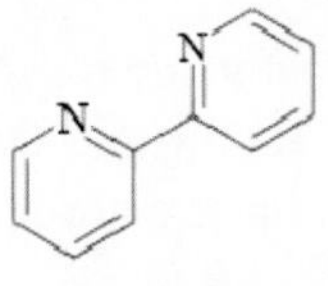

2,2′-bipyridine (bpy)

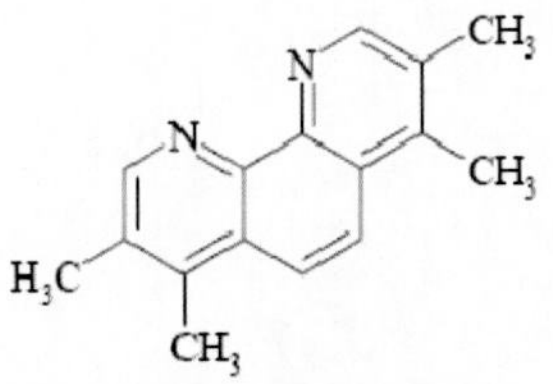

3,4,7,8-tetramethyl-1,10 phenanthroline (tmphen)

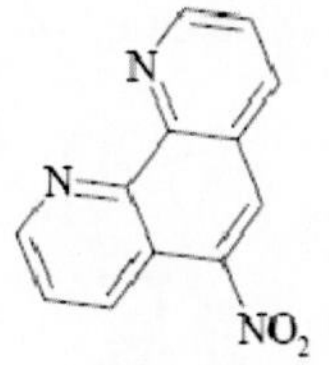

5-Nitro-1,10 phenanthroline (NO_2-phen)

Figure 3. Structural formulae of selected α-diimine and other acceptor ligands and the abbreviations used (in parenthesis).

Marked differences were found between the photochemical and photophysical properties of polymers $\{[(vpy)_2\text{-}vpyRe(CO)_3L]CF_3SO_3\}_{n\sim200}$ and those of the related monomeric complexes $CF_3SO_3[pyRe(CO)_3L]$ (L = phen and bpy). The main cause of these differences is the photogeneration of MLCT excited sates in concentrations that are much larger when $-Re(CO)_3L^+$ chromophores are bound to $(vpy)_{n\sim600}$. This is the photophysical result of Re(I) chromophores being crowded in strands of a polymer instead of being homogeneously distributed through solutions of a $pyRe^I(CO)_3L^+$ complex. The recently communicated association of several hundred strands of $\{[(vpy)_2\text{-}$

vpyRe(CO)$_3$(bpy)]CF$_3$SO$_3$}$_{n\sim200}$ in nearly spherical aggregates also contributes to the crowding of chromophores in small spaces in the solution, where the interaction between excited states becomes appreciable. In figure 4 a fragment of one of these polymers is depicted showing mean distances between Re(I) centers. These mean distances range between 8 and 14 Å in this figure and can be compared with mean distances between chromophores in solutions of pyRe(CO)$_3$L$^+$monomers [17], eq. 1

$$R(\text{in Å}) = \frac{6.5}{\sqrt[3]{C}} \text{ (with C in M units)} \qquad (1)$$

Using a typical value for C in a photochemical experiment (i.e. C ~ 1×10^{-4} M) one obtains R~140 Å. Only at concentrations C > 0.5 M mean distances between monomeric chromophores reach values around 8 Å. Therefore, when chromophores –Re(CO)$_3$(L)$^+$ are attached to the pyridines of (vpy)$_{n\sim600}$ they are spatially confined to small volumes and inter-chromophore distances as low as 8 Å may be achieved at low –Re(CO)$_3$(L)$^+$ concentrations giving rise to new kind of photochemical and photophysical pathways which are not observed in diluted solutions of the respective monomers.

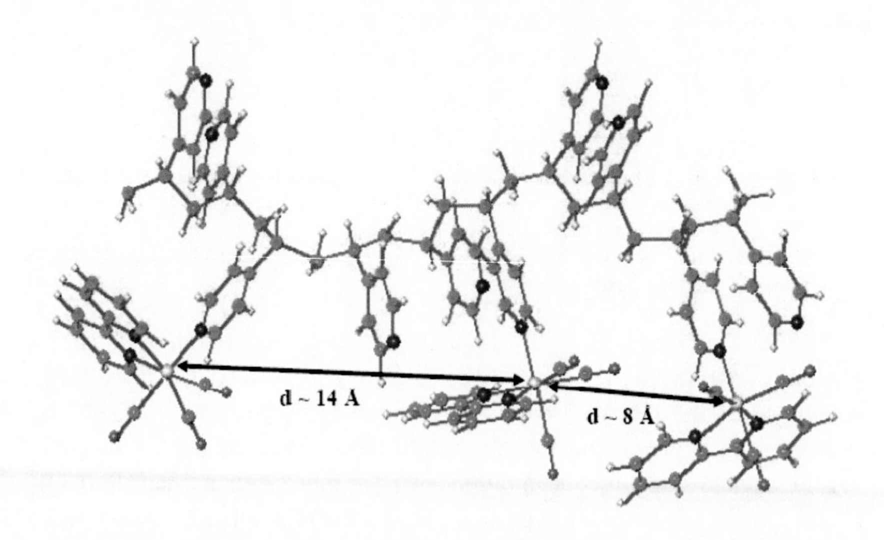

Figure 4. Representation of a fragment (only 3 Re(I) chromophores are represented here) of the polymer [(vpy)$_2$-vpyRe(CO)$_3$(bpy)]$_{n\sim200}$(CF$_3$SO$_3$)$_{n\sim200}$ showing mean inter-chromophore distances.

For instance, the interaction between the Re(I) to phen charge-transfer excited states in close vicinity, ^{3}MLCT, makes the photophysical properties of the pendant chromophores in $\{[(vpy)_2\text{-}vpyRe(CO)_3(phen)]CF_3SO_3\}_{n\sim200}$ deviate from those of $pyRe^I(CO)_3(phen)^+$. The annihilation of two ^{3}MLCT excited states, forms chromophores in the ground state and intraligand, IL, electronic state. The latter being an excited state with a higher energy than the ^{3}MLCT excited state. In contrast to the ^{3}MLCT excited states, the ^{3}IL excited states oxidize organic solvents (e.g., CH_3OH).

In a peralkylated polymer, $\{[(vpy\text{-}CH_3^+)_2\text{-}vpyRe(CO)_3(phen)][CF_3SO_3]_3\}_{n\sim200}$, the 350 nm irradiation of the Re(I) chromophores induces charge-transfer processes between the chromophores. This process is in contrast with the intrastrand energy transfers that occur during the irradiation of $\{[(vpy)_2\text{-}vpyRe(CO)_3(phen)]CF_3SO_3\}_{n\sim200}$. The ^{3}MLCT excited states and charge-separated intermediates in $\{[(vpy\text{-}CH_3^+)_2\text{-}vpyRe(CO)_3(phen)][CF_3SO_3]_3\}_{n\sim200}$ react with neutral (TEOA) or anionic electron donors (e.g., SO_3^{2-} and I^-). The reactions of the anionic electron donors with the excited state of these polyelectrolytes are more efficient than similar reactions with the excited state of $pyRe(CO)_3phen^+$. Similar photochemical and photophysical observations have been made with a polymer $\{[(vpy)_2\text{-}vpyRe(CO)_3(bpy)]CF_3SO_3\}_{n\sim200}$. Since excited-state-excited-state interactions, resulting from the polymer's morphology, have such a large influence on the photophysical and photochemical processes, polymers with more than one type of chromophore are expected to exhibit more varied interactions than those described for polymers $\{[(vpy)_2\text{-}vpyRe(CO)_3(phen)]CF_3SO_3\}_{n\sim200,}$ $\{[(vpy\text{-}CH_3^+)_2\text{-}vpyRe\ (CO)_3(phen)]\ [CF_3SO_3]_3\}_{n\sim200}$ and $\{[(vpy)_2\text{-}vpyRe\ (CO)_3(bpy)]\ CF_3SO_3\}_{n\sim200}$. Also, the specific environments that the strands and aggregates of strands create for the Re(I) chromophores will affect their photochemical properties (e.g.,the lifetime of the MLCT excited states). These phenomena were investigated with mixed polymers containing two different pendants simultaneously bound to $(vpy)_{n\sim600}$ (i.e., $\text{-}Re(CO)_3(phen)^+$ and $\text{-}Re(CO)_3(bpy)^+$) which have similar redox properties, but there are differences in the energy and nature of their excited states. These differences made them suitable compounds to examine these phenomena.

Moreover, the photogeneration of MLCT excited states in close vicinity within a polymer strand makes possible the study of energy transfer processes if donor and acceptor pendants are distributed along the strand. Resonance energy transfer (RET) is a widely prevalent photophysical process through which an electronically excited 'donor' molecule transfers its excitation energy to an 'acceptor' molecule such that the excited state lifetime of the donor decreases. If the donor happens to be a fluorescent molecule RET is referred to as fluorescence resonance energy transfer, FRET. The importance of FRET is ubiquitous. In polymer science, Förster theory is used to study the interface thickness in polymer blends, phase separation and conformational dynamics of polymers. In biological sciences, the technique of FRET is being exploited to design supramolecular systems that can be used to harvest light in artificial photosynthesis as these light-harvesting systems of plants and bacteria involve unidirectional transfer of absorbed radiation energy to the reaction centre via a multistep FRET mechanism. Recent advances in fluorescence resonance energy transfer have led to qualitative and quantitative improvements in the technique, including increased spatial resolution, distance range, and sensitivity. These advances, due largely to new fluorescent dyes, but also to new optical methods and instrumentation, have opened up new biological applications. Besides these, FRET is commonly used in scintillators and chemical sensors.

Polymers with general formula $\{[(vpy)_2vpyRe(CO)_3\ (tmphen)^+]\}_n\{[(vpy)_2vpyRe(CO)_3(NO_2\text{-}phen)^+]\}_m$ were prepared and their morphologies were studied by transmission electron microscopy (TEM). Multiple morphologies of aggregates from these Re^I polymers were obtained by using different solvents. Energy transfer between $MLCT_{Re\rightarrow tmphen}$ and $MLCT_{Re\rightarrow NO2\text{-}phen}$ excited states inside the polymers was evidenced by steady state and time resolved spectroscopy. Current Förster resonance energy transfer theory was successfully applied to energy transfer processes in these polymers.

Chapter 2

SYNTHESIS OF THE POLYMERS

$ClRe(CO)_3L$ complexes are prepared from the commercial (Aldrich) $ClRe(CO)_5$ after reaction with an excess (20%) of the appropriate L ligand in isooctane at reflux under a N_2 atmosphere, eq. 2

$$ClRe(CO)_5 + L\ (\text{excess, }20\%) \xrightarrow{\text{Isooctane, reflux, } N_2} ClRe(CO)_3L + 2CO \qquad (2)$$

Purification of the complex from the excess of L may be realized by dissolution of $ClRe(CO)_3L$ in dichloromethane and precipitation by the slow addition of cold isooctane. $CF_3SO_3Re(CO)_3L$ complexes are prepared and purified by a literature procedure [18] that involves the reaction of $ClRe(CO)_3L$ with $AgCF_3SO_3$, eq. 3

$$ClRe(CO)_3L + CF_3SO_3Ag \xrightarrow{\text{Toluene, reflux, } N_2} CF_3SO_3Re(CO)_3L + AgCl \qquad (3)$$

AgCl may be removed from the product by filtration or by soxhlet extraction of the product if necessary. Reactions of $CF_3SO_3Re(CO)_3L$ with poly-4-(vinylpyridine), average Mw ~ 6 x10^4, were used in the preparations of $[(vpy)_2\text{-}vpyRe(CO)_3L]_{n\sim200}(CF_3SO_3)_{n\sim200}$, according to eq. 4

$$200\ CF_3SO_3Re(CO)_3L + (vpy)_{n\sim600} \xrightarrow{\text{Dichloromethane, reflux, } N_2} [(vpy)_2\text{-}vpyRe(CO)_3L]_{n\sim200}(CF_3SO_3)_{n\sim200} \quad (4)$$

In the synthesis of the polymers, to a solution containing $(vpy)_{n\sim600}$ (5.6×10^{-7} mol) in 20 cm^3 of CH_2Cl_2 was slowly added by stirring 1.1×10^{-4} mol of $CF_3SO_3Re(CO)_3L$ in 50 cm^3 of CH_2Cl_2. This stoichiometric relationship makes 200 $CF_3SO_3Re(CO)_3L$ react with a similar number of pyridine groups of the ca. 600 present in the polymer. A yellow solid precipitated during the 9 h that the solution was refluxed under a blanket of N_2. The mixture was rotoevaporated to dryness; the resulting solid was redisolved in the minimum volume of CH_3CN, and the polymer was precipitated by the slow addition of ethyl ether. In the synthesis of mixed polymers, the reactions with $(vpy)_{n\sim600}$ were carried out using CH_2Cl_2 solutions that contained both $CF_3SO_3Re(CO)_3L_1$ and $CF_3SO_3Re(CO)_3L_2$ but maintaining the total number of chromophores in a ratio 200/600 to the total number of pyridine units in the final polymer. By this way the following polymers were synthesized:

(I): $[(vpy)_2\text{-}vpyRe(CO)_3(phen)]_{n\sim200}(CF_3SO_3)_{n\sim200}$

(II): $[(vpy)_2\text{-}vpyRe(CO)_3(bpy)]_{n\sim200}(CF_3SO_3)_{n\sim200}$

(III): $[(vpy)_{m\sim338}\text{-}(vpy\text{-}Re(CO)_3(bpy))_{n\sim131}(vpy\text{-}Re(CO)_3(phen)_{p\sim131}]$ $(CF_3SO_3)_{n+p\sim262}$

(IV): $[(vpy)_{m\sim250}\text{-}(vpy\text{-}Re(CO)_3(bpy))_{n\sim200}(vpy\text{-}Re(CO)_3(phen)_{p\sim150}]$ $(CF_3SO_3)_{n+p\sim350}$

(V): $[(vpy)_2\text{-}vpyRe(CO)_3(tmphen)]_{n\sim200}(CF_3SO_3)_{n\sim200}$

(VI): $[(vpy)_2\text{-}vpyRe(CO)_3(NO_2\text{-}phen)]_{n\sim200}(CF_3SO_3)_{n\sim200}$

(VII): $[(vpy)_{m\sim400}\text{-}(vpy\text{-}Re(CO)_3(tmphen))_{n\sim180}$ $(vpy\text{-}Re(CO)_3(NO_2\text{-}phen)_{p\sim20}](CF_3SO_3)_{n+p\sim200}$

(VIII): $[(vpy)_{m\sim400}\text{-}(vpy\text{-}Re(CO)_3(tmphen))_{n\sim150}$ $(vpy\text{-}Re(CO)_3(NO_2\text{-}phen)_{p\sim50}](CF_3SO_3)_{n+p\sim200}$

(IX): $[(vpy)_{m\sim400}\text{-}(vpy\text{-}Re(CO)_3(tmphen))_{n\sim100}$
$(vpy\text{-}Re(CO)_3(NO_2\text{-}phen)_{p\sim100}](CF_3SO_3)_{n+p\sim200}$

(X) : $[(vpy\text{-}CH_3^+)_2\text{-}vpyRe(CO)_3(phen)]_{n\sim200}(CF_3SO_3)_{n\sim600}$

Where the polymer **(X),** i.e. $[(vpy\text{-}CH_3^+)_2\text{-}vpyRe(CO)_3(phen)]_{n\sim200}$ $(CF_3SO_3)_{n\sim600}$ was obtained by peralkylation of the free pyridines of polymer **(I)**, eqs. 5-6

$$[(vpy)_2\text{-}vpyRe(CO)_3(phen)]_{n\sim200}(CF_3SO_3)_{n\sim200} \xrightarrow{ICH_3} \quad (5)$$

$$[(vpy\text{-}CH_3^+)_2\text{-}vpyRe(CO)_3(phen)]_{n\sim200}(CF_3SO_3)_{n\sim200}(I)_{m\sim400}$$

$$[(vpy\text{-}CH_3^+)_2\text{-}vpyRe(CO)_3(phen)]_{n\sim200}(CF_3SO_3)_{n\sim200}(I)_{m\sim400} \xrightarrow[-AgI]{AgCF_3SO_3} \quad (6)$$

$$[(vpy\text{-}CH_3^+)_2\text{-}vpyRe(CO)_3(phen)]_{n\sim200}(CF_3SO_3)_{n\sim600}$$

In the following sections the morphological, photophysical and photochemical properties of polymers **(I)-(X)** will be discussed.

Chapter 3

SOLVENT EFFECTS ON MORPHOLOGIES: NANOAGGREGATION

The morphology of polymer **(II)** was studied by transmission electron microscopy (TEM). The polymer films were obtained by room temperature solvent evaporation of its acetonitrile solutions. When taking photos, the polymer films were not stained with any chemicals, and the contrast of the image in the TEM photos can only originate from the rhenium complexes incorporated to the polymers. The Re(I) complexes in the polymer aggregate and form isolated nanodomains that are dispersed in the polymer matrix film. The dimensions of the nanodomains are between 90 and 430 nm and are mainly spherical in shape. Similar results were obtained from dynamic and static light scattering measurements on polymers **(II), (III)** and **(IV)** indicating that the nanoaggregates also exist in solution [19]. The morphologies of polymers **(V)-(IX)** were also studied by TEM. Multiple morphologies of aggregates from these Re(I) polymers were obtained by using different solvents. TEM images of acetonitrile and dichloromethane- cast films of the polymers are shown in figure 5 and figure 6, respectively. Morphologies of the polymers differed when the cast films were obtained either from acetonitrile or dichloromethane solutions. When the solvent was acetonitrile, figure 5, in the solid phase of polymer **(VI)**, the Re(I) complexes aggregate and form isolated nanodomains that are dispersed in the $(vpy)_{n\sim600}$ backbone. The dimensions of the aggregates are between 80 and 160 nm and are mainly spherical in shape. However, polymer **(V)** does not aggregate to form large nanodomains and only small spherical objects with diameter between 5 and 30 nm, are observed. In polymers **(VII)-(IX)**, TEM images suggest the formation of aggregates of increasing size from **(VII)** to **(IX)**. It should be noted that the dimensions of the nanodomains are considerably larger than the full stretch length of the polymers. As a result, it is likely that they contain more than one layer of polymers. The situation is completely different in dichloromethane. Figure 6 shows TEM images of dichloromethane-cast films of polymers **(V)-(IX)**. Vesicles were obtained for polymers **(V), (VII)**, and **(VIII)**. The vesicular nature is evidenced by a higher transmission in the center of the aggregates than around their periphery in the TEM pictures. The sizes of the vesicles are very polydisperse with outer diameters ranging from 140 nm to large compound vesicles with diameters up to 1.4 μm. More interestingly, polymer **(VI)** formed branched tubular structures intertwined in a net.

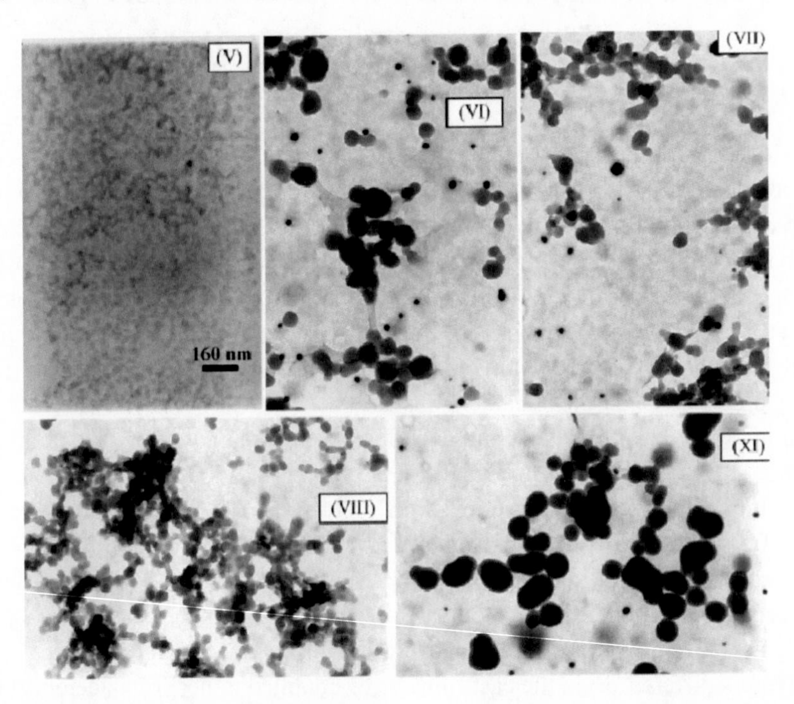

Figure 5. Acetonitrile-cast films of polymers (V), (VI), (VII), (VII) and (XI). The same scale bar apply to all the films in the figure.

The morphologies shown for polymer **(IX)** are the intermediate shape of vesicles and tubules. It is interesting to note that micellar-like aggregates have been previously also observed from TEM images of acetonitrile-cast films of polymer **(III)** and **(IV)** [20]. We can rationalize the solvent effect upon aggregation of Re(I)-$(vpy)_{n\sim600}$ polymers as follows. For instance, $(vpy)_{n\sim600}$ is nearly insoluble in acetonitrile, but this solvent is a good one for the Re(I)-$(vpy)_{n\sim600}$ polymers. Then, it is plausible to imagine that the inner core of the micelles present in acetonitrile will be formed mainly by the free pyridines of the Re(I)-$(vpy)_{n\sim600}$ polymers and the outer part will be constituted mainly by the solvated Re(I) pendants. The situation is reversed in dichloromethane as this solvent is a good one for $(vpy)_{n\sim600}$. Moreover, polymers **(I)** and **(II)** cannot be dissolved in that solvent while solubility of polymers **(V)-(IX)** is considerably lower in dichloromethane than in acetonitrile. In vesicles, however, a pool of dichloromethane molecules are solvating the uncoordinated pyridines of the polymer in the inner and outer region of the vesicle while the Re(I) pendants mainly remain inside the membrane of the vesicle.

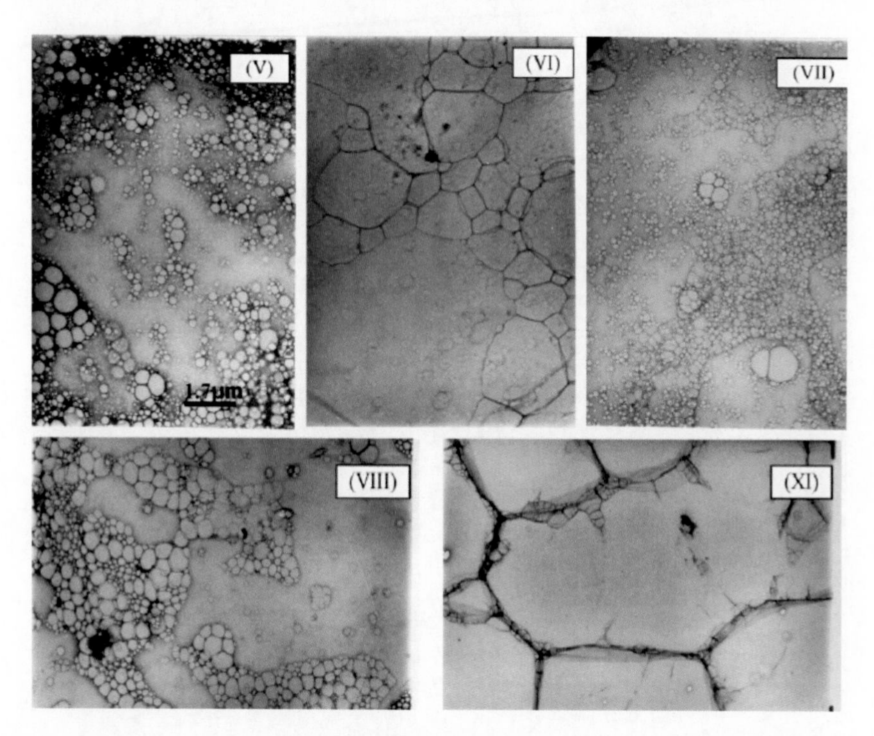

Figure 6. Dichloromethane-cast films of polymers (V), (VI), (VII), (VII) and (XI). The same scale bar apply to all the films in the figure.

Chapter 4

PHOTOPHYSICAL PROPERTIES OF POLYMER (I)

The polymer and the monomer $[pyRe(CO)_3phen]CF_3SO_3$ exhibited UV-vis spectra with similar features, but the extinction coefficients of the polymer, by comparison to the extinction coefficients of the monomer, corresponded to ~200 chromophores, $-Re(CO)_3phen^+$, per formula weight of polymer. This load of Re(I) pendants was in good agreement with a calculation from the elemental analysis and with the structures shown in figure 2. The emission spectra of $pyRe(CO)_3phen^+$ and $\{(vpy)_2\text{-}vpyRe(CO)_3phen^+\}_{n\sim200}$ in deaerated CH_3CN exhibited bands respectively centered at 560 and 565 nm with identical band shapes. Also, 351 nm laser flash irradiations of the monomer and polymer in deaerated CH_3CN produced nearly identical transient absorption spectra respectively assigned to the MLCT excited states. The spectra of the excited state, λ_{max} ~460 nm and λ_{max} ~750 nm, did not change with the number n_{hv} of 351 nm photons absorbed by either $pyRe(CO)_3phen^+$ or $\{(vpy)_2\text{-}vpyRe(CO)_3phen^+\}_{n\sim200}$. By contrast to this invariance of the MLCT spectra with n_{hv}, the quantum yield of the MLCT excited state, ϕ_{MLCT}, and its lifetime showed marked dependences on n_{hv} and the photogenerated concentration of MLCT excited state. For the monomer $pyRe(CO)_3phen^+$, although the quantum yield was nearly independent of n_{hv} for n_{hv} values that were below 20% of the total Re(I) concentration, it decreased monotonically above that limit. Similar experiments with solutions of the polymer$\{(vpy)_2\text{-}vpyRe(CO)_3phen^+\}_{n\sim200}$ in CH_3CN or $MeOH/H_2O$ mixed solvent showed that ϕ_{MLCT} decreased with n_{hv} and reached the same value of the monomer in the limit $n_{hv} \rightarrow 0$ (see figure 7). The disappearance of the MLCT excited states in the monomer and the polymer occurred with different reaction kinetics. In experiments with the polymer, the decay of the excited state exhibited more marked second order behavior than the monomer. In fact, linear inverse plots of the reciprocal of the change in optical density, ΔA^{-1} vs time and the linear dependence of $t_{1/2}$ on the reciprocal of the MLCT concentration, (figure 8) demonstrated that the MLCT decay was kinetically of a second order on the MLCT concentration for $n_{h\nu}$ values larger than 17% of the total Re(I) concentration. A similar study of the MLCT decay kinetics with the monomer established that traces were well fitted to a single exponential and the lifetime exhibited no dependence on MLCT concentration until $n_{h\nu}$ corresponded to 30% of the Re(I) concentration. These experimental observations showed that the process whose rate was kinetically of a second order in MLCT concentration contributed less to the excited state decay in the monomer [21].

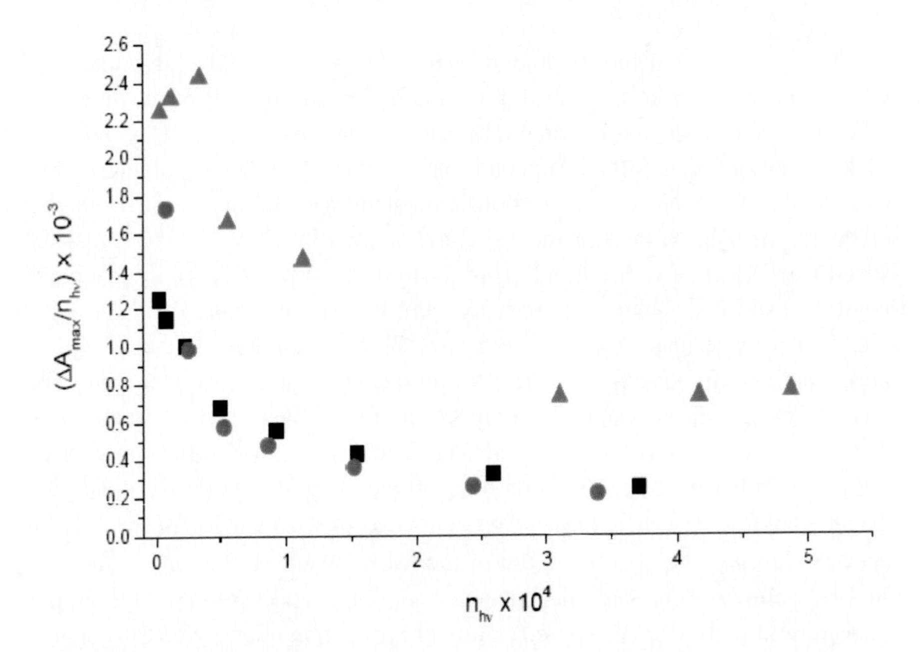

Figure 7. Dependence of the MLCT excited-state quantum yield, ϕ_{MLCT} (assumed proportional to $\Delta A_{max}/n_{h\nu}$) , on $n_{h\nu}$ in laser flash photolysis (351 nm) of the Re(I) monomer $pyRe(CO)_3phen^+$ and the Re(I) polymer $\{(vpy)_2\text{-}vpyRe(CO)_3phen^+\}_{n\sim200}$. Solutions of the Re(I) complexes in deaereated CH_3CN were used with $pyRe(CO)_3phen^+$ (▲) and $\{(vpy)_2\text{-}vpyRe(CO)_3phen^+\}_{n\sim200}$ (■). Solutions in deaereated $MeOH/H_2O$ (1:4) of the Re(I) polymer were used in (●) experiments.

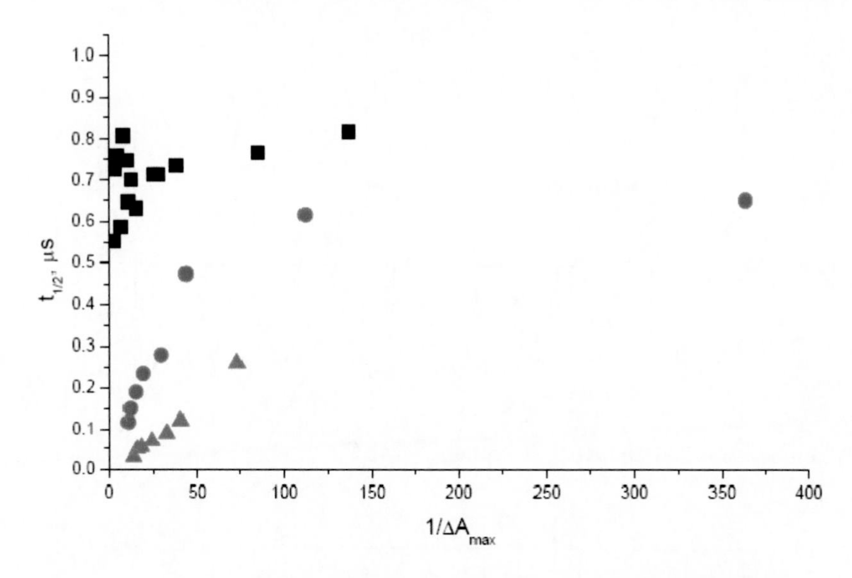

Figure 8. Dependence of MLCT state $t_{1/2}$ on $1/\Delta A_{max}$ for $pyRe(CO)_3phen^+$ (■, in CH_3CN) and $\{(vpy)_2\text{-}vpyRe(CO)_3phen^+\}_{n\sim200}$ (●, in CH_3CN and ▲, in MeOH-H_2O).

Relative to $pyRe(CO)_3phen^+$, the incorporation of the -$Re(CO)_3phen^+$ chromophore in a poly-4-(vinylpyridine) polymer backbone has little effect on the photophysical processes of the MLCT excited state when these complexes are irradiated with low photonic fluxes, i.e., steady-state photolysis or laser irradiations with $n_{hv} < 5$ mJ/pulse $\approx 5x10^{-6}$ Einstein L^{-1} $pulse^{-1}$. This experimental observation and the similarity of the monomer and polymer absorption spectra suggest that electronic interactions between Re(I) chromophores in the polymer are negligible. A rigid rod conformation of the polymer backbone that supports maximum separation between cationic pendants accounts for these spectroscopic and photophysical properties. By contrast to the irradiations with low photonic fluxes, the photophysics of the MLCT in the Re(I) monomer is different of the one observed with the Re(I) polymer when irradiations are carried out with laser powers $n_{hv} \geq 5$ mJ/pulse. In the Re(I) monomer, the quantum yield of the MLCT excited state, ϕ_{MLCT}, remains constant until there is a significant depletion of the ground-state population, i.e., $n_{hv} \leq 3.5x10^{-5}$ Einstein L^{-1} $pulse^{-1}$ in figure 7. The decrease of ϕ_{MLCT} with n_{hv} occurs earlier than expected for the depletion of the ground state in the Re(I) polymer. However, absorption of light by MLCT excited states in the polymer to form intraligand excited states, IL, accounts for the functional dependence of ϕ_{MLCT} on n_{hv}, eqs. 7, 8.

$$\text{---}\left[\text{py}\ \ \text{py}(\text{Re(CO)}_3\text{phen}^+)\ \ \text{py}\right]_{n\sim200}\text{---} \xrightarrow{h\nu}$$

$$\text{---}\left[\text{py}\ \ \text{py}(\text{Re(CO)}_3\text{phen}^+)\ \ \text{py}\right]_x\text{---}\left[\text{py}\ \ \text{py}(\text{Re}^*\text{(CO)}_3\text{phen}^+)\ \ \text{py}\right]_y\text{---} \quad (7)$$

Where x + y ~ 200 and $Re^*(CO)_3phen^+$ = MLCT

$$\text{---}\left[\text{py}\ \ \text{py}(\text{Re(CO)}_3\text{phen}^+)\ \ \text{py}\right]_x\text{---}\left[\text{py}\ \ \text{py}(\text{Re}^*\text{(CO)}_3\text{phen}^+)\ \ \text{py}\right]_y\text{---} \xrightarrow{h\nu} \quad (8)$$

$$\text{---}\left[\text{py}\ \ \text{py}(\text{Re(CO)}_3\text{phen}^+)\ \ \text{py}\right]_p\text{---}\left[\text{py}\ \ \text{py}(\text{Re}^*\text{(CO)}_3\text{phen}^+)\ \ \text{py}\right]_q\text{---}\left[\text{py}\ \ \text{py}(\text{Re(CO)}_3\text{(phen}^*)^+)\ \ \text{py}\right]_n\text{---}$$

Where p + q + n ~ 200 and $Re(CO)_3(phen^*)^+$ = IL

$$\text{---}\left[\text{py}\ \ \text{py}(\text{Re(CO)}_3\text{phen}^+)\ \ \text{py}\right]_p\text{---}\left[\text{py}\ \ \text{py}(\text{Re}^*\text{(CO)}_3\text{phen}^+)\ \ \text{py}\right]_q\text{---}\left[\text{py}\ \ \text{py}(\text{Re(CO)}_3\text{(phen}^*)^+)\ \ \text{py}\right]_n\text{---}$$

$$\longrightarrow \text{---}\left[\text{py}\ \ \text{py}(\text{Re(CO)}_3\text{phen}^+)\ \ \text{py}\right]_{p+n}\text{---}\left[\text{py}\ \ \text{py}(\text{Re}^*\text{(CO)}_3\text{phen}^+)\ \ \text{py}\right]_q\text{---} \quad (9)$$

Because the IL excited state is shorter lived than the MLCT excited state, the decay represented in eq. 9 is faster than the instrument's 20 ns response, and it was manifested only by a photogenerated concentration of MLCT smaller than the expected one.

A fast intrastrand annihilation of MLCT excited states also provides a good rationale for the functional dependence of ϕ_{MLCT} on $n_{h\nu}$. In this mechanism, the rapid curvature of ϕ_{MLCT} with $n_{h\nu}$ in figure 7 requires that a large fraction of photogenerated excited states vanishes within the 20 ns laser pulse. If excited Re chromophores are close neighbors and have the right spatial orientation within the strand of polymer, they may undergo a fast annihilation within that period of time. This fast annihilation process can create IL excited states that are placed at higher energies and that are more reactive than the parent MLCT excited state.

Excited chromophores that are disfavored by reason of their position for a fast annihilation or do not undergo a secondary photolysis will be observed at times longer than the 20 ns laser irradiation. Flash photolysis shows that the rate of decay of this remanent excited state population is kinetically of a second order in the overall MLCT concentration. Because diffusive motions of the polyelectrolyte are much slower than those of the observed decay of excited states, the second-order kinetics indicates that mechanisms by which the energy moves through the strand of polymer are available. Various mechanisms of intramolecular energy transfer have been proposed for organic polymers [22]. In a mechanism, energy hopping in the strand may form pairs of excited chromophores that undergo annihilations. The mechanism is supported by experimental observations with $\{(vpy)_2\text{-}pyRe(CO)_3phen^+\}_{n\sim200}$ after uncomplexed pendant pyridine groups are methylated. Quaternization of the pyridine groups enhanced electron transfer through the strand of the polymer. The process could involve excited states of the uncomplexed pyridine pendants. It is also possible that energy can be transferred between remotely placed excited chromophores. These events will leave, therefore, one chromophore in the ground state and the other in an upper intraligand excited state, IL.

Chapter 5

LASER POWER, THERMAL AND SOLVENT EFFECTS ON THE PHOTOPHYSICAL PROPERTIES OF POLYMER (II)

The emission spectra of the monomer and the polymer with L= bpy showed distinctive behavior in solvent mixtures . For instance the emission spectra of $CF_3SO_3[pyRe(CO)_3bpy]$ and$\{[(vpy)_2\text{-}vpyRe(CO)_3bpy]\ CF_3SO_3\}_{n\sim200}$ in deoxygenated DMSO-CH_3CN (1: 9 v/v) mixtures at room temperature exhibited unstructured bands, respectively centered at 575 and 577 nm with similar band shapes. When the solvent was DMSO-CH_2Cl_2 (1 : 9 v/v) the emission maximum shifted, for both monomer and polymer, to 569 nm. However, when the solvent was DMSO-water (1 : 9 v/v) mixtures, the emission maximum for the monomer, at 574 nm, was nearly coincident with that in DMSO-CH_3CN (1: 9 v/v) while in the case of the polymer emission maximum (at 561 nm) shifted to lower wavelengths. Emission decays in the mixed solvents DMSO-CH_3CN, DMSO-CH_2Cl_2 and DMSO–water were monoexponential for $CF_3SO_3[pyRe(CO)_3bpy]$. However, for the polymer $\{[(vpy)_2\text{-}vpyRe(CO)_3bpy]\ CF_3SO_3\}_{n\sim200}$, monoexponential emission decays were only observed with DMSO-CH_3CN mixtures, while in DMSO–water and DMSO–CH_2Cl_2 mixtures the emission decay became biexponential [23]. Table 1 summarizes steady state luminescence maxima and time resolved luminescence and transient absorbance lifetimes for $CF_3SO_3[pyRe(CO)_3bpy]$ and$\{[(vpy)_2\text{-}vpyRe(CO)_3bpy]\ CF_3SO_3\}_{n\sim200}$ in different experimental conditions.

Table 1. Emission maxima and emission and transient absorption lifetimes for $CF_3SO_3[pyRe(CO)_3(2,2'bipy)]$ (Monomer) and $\{[(vpy)_2\text{-}vpyRe(CO)_3(2,2'bipy)]\ CF_3SO_3\}_{200}$ (Polymer) in different experimental conditions

Compound	Solvent	λ_{em} / nm	τ_{em} / ns	τ_{abs} / ns	Conditions
Monomer	CH_3CN	-	260	282	λ_{exc}=355 nm, RT[a]
Monomer	MeOH	-	218	227	λ_{exc}=355 nm, RT[a]
Monomer	CH_3CN		-	225	λ_{exc}=351 nm, RT[b]
Monomer	CH_3CN	-	245	-	λ_{exc}=337 nm, RT[c]
Monomer	DMSO-CH_3CN	575	213	-	λ_{exc}=337 nm, RT[c]
Monomer	DMSO-CH_2Cl_2	569	316	-	λ_{exc}=337 nm, RT[c]
Monomer	DMSO-water	574	133	-	λ_{exc}=337 nm, RT[c]
Polymer	CH_3CN	-	226, 53	187, <10	λ_{exc}=355 nm, RT[a]
Polymer	MeOH	-	201, 67	185	λ_{exc}=355 nm, RT[a]
Polymer	CH_3CN	-	-	184, 35	λ_{exc}=351 nm, RT[b]
Polymer	CH_3CN	-	203	-	λ_{exc}=337 nm, RT[c]
Polymer	DMSO-CH_3CN	577	137	-	λ_{exc}=337 nm, RT[c]
Polymer	DMSO-CH_2Cl_2	569	143, 31	-	λ_{exc}=337 nm, RT[c]
Polymer	DMSO-water	561	440, 83	-	λ_{exc}=337 nm, RT[c]

[a]Nd-YAG laser [b]Excimer laser [c]Nitrogen laser.

The effect of temperature between 0°C and 65°C on the emission lifetime was studied irradiating at 337 nm (2 mJ/pulse) N_2-deoxygenated CH_3CN and/or CH_3CN-water (1:4) solutions of the monomer and the polymer. In CH_3CN solutions emission decays were monoexponential for both the monomer and the polymer within the whole temperature range of study. Moreover, the temperature dependence of emission lifetime is nearly the same for the two compounds. However, when the solvent was CH_3CN-water mixture, results were very different for monomer than for polymer. The monomer emission decay remains monoexponential between 0°C and 65°C showing nearly the same slope, i.e. $\partial\ln(1/\tau)/\partial T$, than in the case of CH_3CN solutions. However, the polymer showed a biexponential behavior at temperatures below 15°C, with a higher $\partial\ln(1/\tau)/\partial T$ showing a much more marked dependence of τ on temperatures between 15°C and 0°C than between 15°C and 65°C [23].

As it has previously been observed with -Re(CO)$_3$phen pendants in a (vpy)$_{n\sim600}$ backbone, the incorporation of the 200 -Re(CO)$_3$bpy chromophores to (vpy)$_{n\sim600}$ to form the polymer {[(vpy)$_2$-vpyRe(CO)$_3$bpy] CF$_3$SO$_3$}$_{200}$ has little effect, relative to the monomer CF$_3$SO$_3$[pyRe(CO)$_3$bpy], on the photophysical processes of the MLCT excited state when these complexes are irradiated with low photonic fluxes, i.e. steady state or laser flash irradiations with $n_{h\nu} \leq 2$ mJ/pulse. In fact, emission decay lifetimes of both CF$_3$SO$_3$[pyRe(CO)$_3$bpy] and {[(vpy)$_2$-vpyRe(CO)$_3$bpy] CF$_3$SO$_3$}$_{200}$ in CH$_3$CN, calculated from 337 nm flash excitations experiments, are monoexponential and very similar. Though a longer lifetime for the monomer (τ_{em} = 245 ns) than for the polymer (τ_{em} = 203 ns) could be associated to the availability of new deactivation pathways for the MLCT in the polymer due to vibration modes present in the poly-vinylpyridine backbone. This experimental observation and the similarity of the monomer and the polymer absorption and emission spectra suggest that electronic interactions between Re(I) chromophores in the polymer are negligible in solvents like CH$_3$CN or DMSO-CH$_3$CN mixtures. A rigid rod conformation of the polymer backbone that support maximun separation between cationic pendants could account for these experimental observations. However, this would not be the case for solvents of very low or very high polarity, as in DMSO-CH$_2$Cl$_2$ and DMSO-water mixtures respectively. Although the shape and maxima of emission spectra in DMSO-CH$_2$Cl$_2$ are nearly the same for the monomer and the polymer, the polymer experiences a biexponential emission being τ_1 = 143 ns and τ_2 = 31 ns respectively. Those lifetimes are considerably shorter than that for the monomer, τ = 316 ns. Moreover, the polymer emission spectrum in DMSO-water is blue-shifted (~ 13 nm) compared to the one of the monomer and the emission decay is also biexponential. A coil structure enhancing the interaction between vicinal MLCTs and bringing them to an environment where the solvent molecules are mostly excluded, rather than a rigid rod one, might be responsible of this distinct photophysical behavior.

This proposition is well supported by the formation of aggregates in a similar polymer, i.e. polystyrene-block-poly(4-vinylpyridine) (PS-b-PVP) functionalized with pendants -Re(CO)$_3$(bpy)$^+$ groups [24]. In fact, addition of MeOH or Toluene to a CH$_2$Cl$_2$ solution of the PS-b-PVP- Re(CO)$_3$(bpy)$^+$ polymer resulted in the formation of micelles. Generally, block copolymers exhibit various morphological properties in different solvent systems because of the nature of the different blocks. They are able to form a number of micelle structures such as spherical, lamellar, and cylindrical in shape. Two PS-b-PVP- Re(CO)$_3$(bpy)$^+$ polymers with different Re(I) content exhibited the formation of rod-like and spherical in shape micelles respectively, as shown by transmission electron microscopy and light scattering experiments. For instance, in these polymers, the Re-PVP blocks constitute the outer part of the sphere and the PS blocks remain inside the spherical micelle. It is noteworthy that in despite of the poly-4-vinylpyridine is quite soluble in dichloromethane, the polyelectrolyte $\{[(vpy)_2\text{-}vpyRe(CO)_3bpy]\ CF_3SO_3\}_{n\sim200}$ cannot be dissolved in that solvent. This fact indicates that solvents of low polarity like dichloromethane can no longer sustain the large number of positive charges (~200 per molecule) present in the polyelectrolyte. With the $\{[(vpy)_2\text{-}vpyRe(CO)_3bpy]\ CF_3SO_3\}_{n\sim200}$ polymer, however, no precipitation occurs in DMSO-water or DMSO- CH_2Cl_2 mixed solvents. However, under conditions of highly polar or highly non polar solvents, spherical structures, though still soluble in the mixed solvents, might be occurring. Although the distribution of pendant $-Re(CO)_3(bpy)^+$ groups in the $\{[(vpy)_2\text{-}vpyRe(CO)_3bpy]\ CF_3SO_3\}_{200}$ polymer should be statistical, i.e. ~200 $-Re(CO)_3(bpy)^+$ groups distributed at random among ~600 pyridines, one can imagine non-statistical regions within the polymer where there are blocks of uncoordinated pyridines and blocks of coordinated pyridines, resembling the PS and PVP blocks in the PS-b-PVP-$Re(CO)_3(bpy)^+$ polymers.

Temperature dependence of emission lifetimes in high polar mixtures (CH_3CN-water) also show a biexponential behaviour for the polymer at temperatures below 15^0C. Above that temperature its behaviour is monoexponential. Here thermal activation seems responsible for the transition between the coil and the rigid rod structures, the coil structure occurring in highly polar (or in highly non-polar) solvents at temperatures below 15^0C and the rigid rod structure in solvents like CH_3CN and at temperatures above 15^0C. In this regard, in the related poly (3-methyl-4-vinylpyridine) the optical activity of the polymer solution has been ascribed to a helical conformation. This optical activity is lost in solution at -4^0C probably due to a conformational transition [25].

Laser flash irradiations with $n_{hv} \geq 12$ mJ/pulse at $\lambda_{ex} = 351$ nm or 355 nm show a very different photophysical behavior for $\{[(vpy)_2\text{-}vpyRe(CO)_3bpy]\ CF_3SO_3\}_{200}$ and $CF_3SO_3[pyRe(CO)_3bpy]$, respectively. The quantum yield of the MLCT formation is nearly three times higher for the monomer than for the polymer. Besides, emission decay for the monomer is monoexponential with a lifetime that is nearly the same than that observed at irradiations with $n_{hv} \leq 2$ mJ/pulse at $\lambda_{ex} = 337$ nm. However, for the polymer, emission decay could only be fitted with two exponentials and the transient generated also experiences a biexponential decay. A fast intrastrand annihilation of MLCT excited states can provide a good explanation for the lower amount of MLCT excited states observed in flash photolysis experiments with the polymer compared to that of the monomer. This mechanism requires that a large fraction of the photogenerated excited states vanishes within the 20 ns laser pulse. If excited -$Re(CO)_3$bpy chromophores are close neighbours and have the correct spatial orientation within the polymer strand, they may undergo a fast annihilation within that period of time. However, using high intensity pulses, a secondary photolysis of the MLCT in the polymer to form intraligand excited states, as discussed above for $\{[(vpy)_2\text{-}vpyRe(CO)_3phen]\ CF_3SO_3\}_{200}$ polymer, can not be discarded in this rationalization. Excited chromophores that are disfavored by reason of their position for a fast annihilation or do not undergo secondary photolysis will be observed decaying at times longer than 20 ns. The number of these excited chromophores will be higher in flash photolysis experiments with high laser powers (i.e. λ_{ex} = 351 nm or 355 nm) than in flash photolysis experiments with the low power N_2 laser (λ_{ex} = 337 nm) explaining why 351 nm excitation produces a transient that decays biexponentially and the transient produced after 337 nm excitation decays by a first order kinetics. In the related $\{[(vpy)_2\text{-}vpyRe(CO)_3phen]\ CF_3SO_3\}_{200}$ polymer a second order process was observed in addition to the MLCT first order decay. It is worthy to note (see Table 1) that the longer luminescence lifetime observed after 355 nm or 351 nm excitation is nearly the same to that observed after 337 nm photolysis. The shorter lifetime in 351 nm experiments could be ascribed to the second order processes mentioned before.

PHOTOCHEMICAL PROPERTIES OF POLYMER (I)

Redox reactions of the MLCT excited state, in addition to the radiative and radiationless relaxations observed in deaerated CH_3CN, were detected with solutions of polymer **(I)** in MeOH or mixed solvents containing MeOH. In addition to a faster rate of decay in MeOH or MeOH-water mixtures, the spectrum recorded after the decay of the excited $\{(vpy)_2\text{-}vpyRe(CO)_3phen^+\}_{n\sim200}$ presented an absorption band with λ_{max} = 550 nm. This 550 nm absorption band is characteristic of the reduced Re(I) complex with a coordinated $phen^{\bullet-}$ chromophore [21]. The effect of the solvent on the oxidative quenching of the MLCT was investigated with solutions of $\{(vpy)_2\text{-}vpyRe(CO)_3phen^+\}_{n\sim200}$ in deaerated CH_3CN or MeOH that also contained methyl viologen (MV^{2+}). The spectrum of reduced methyl viologen, MV^+, observed after the decay of the MLCT excited state showed that the yield of MV^{+} in MeOH solutions was five times larger than in CH_3CN. This experimental observation and the lifetimes in figure 7 gave a ratio of the quenching rate constant in MeOH to the rate constant in CH_3CN that was between 6 and 7. However, the quenching rate constants measured from the optical density changes at 450 nm, k= $4.3x10^{8}$ M^{-1} s^{-1} in CH_3CN and k= $2.1x10^{9}$ M^{-1} s^{-1} in MeOH, differed by a factor of 5. The rate constant for the monomer's MLCT quenching was measured in flash photolysis of $pyRe(CO)_3phen^{+}$ in CH_3CN containing MV^{2+} and with the same n_{hv} used in the experiments with $\{(vpy)_2\text{-}vpyRe(CO)_3phen^{+}\}_{n\sim200}$. The rate constant measured from the decay of the optical density at 450 nm, k = $1.3x10^{9}$ M^{-1} s^{-1}, was only three times larger than the one measured with the polymer [21].

The reductive quenching of the MLCT excited state was investigated by flash photolysis of $pyRe(CO)_3phen^{+}$ and $\{(vpy)_2\text{-}vpyRe(CO)_3phen^{+}\}_{n\sim200}$ in deaerated CH_3CN containing triethanolamine (TEOA) in concentrations equal to or larger than 0.073 M. Identical concentrations of the Re(I) chromophore, [Re(I)] = 1.7 x 10^{-4} M] , were used in solutions of the monomer and the polymer. Transient spectra, λ_{max} = 420, 550 nm and $\lambda_{max} > 750$ nm, recorded by flash photolysis, were assigned to Re(I) species with coordinated $phen^{\bullet -}$. In a 20 μs time scale, processes similar to those described in the literature reports, i.e., the additional formation of $pyRe(CO)_3(phen^{\bullet -})$ due to reactions of excess Re(I) complex with radicals from the TEOA oxidation, followed the photogeneration of $pyRe(CO)_3(phen^{\bullet -})$, figure 9. The disappearance of the $pyRe(CO)_3(phen^{\bullet -})$ spectrum at times longer than 0.1 s took place with a rate that was kinetically of a second order in accordance with the expected disproportionation of the radicals. Reduction of the polymer's excited states was effected in less than 30 ns by TEOA. This process was followed at times t > 25 μs by a growth of the - $Re(CO)_3(phen^{\bullet -})$ spectrum that showed an additional reduction of -$Re(CO)_3phen^{+}$ groups in direct proportion to the TEOA concentration, figure 10. Further spectral changes took place in a 200 μs time scale, namely after the formation of -$Re(CO)_3(phen^{\bullet -})$. Because the spectra recorded between 200 and 300 μs and transient spectra generated in the disproportionation of pyRe(CO)3($phen^{\bullet -}$) exhibited similar spectral features, it was related to the formation of two electron-reduced groups, -$Re(CO)_3(phenH_2)$, in the polymer [21]. It must be noted that the process leading to the formation of -$Re(CO)_3(phenH_2)$ in the polymer was 10 times faster than that of the disproportionation of the monomer's Re(I) ligand-radical and that oscillographic traces were fitted to a single exponential with a rate constant k = (1.10 ± 0.05)x10^4 s^{-1} instead of the second-order kinetics observed with the monomer. The same experiments were conducted in a MeOH/H_2O, 20% v/v, mixed-solvent. Transient spectra of the monomer and polymer Re(I) ligand-radicals disappeared by processes that were kinetically of a second order in transient concentration. In either solvent, CH_3CN or MeOH/H_2O, the decay of the transient spectra was not complete, and the residual spectrum, similar to one recorded for the polymer's -Re$(CO)_3$(phen$^{\bullet-}$), remained stable for several minutes after the irradiation [21]. To a certain extent, medium effects were observed in the excited state redox quenching by methyl viologen or TEOA.

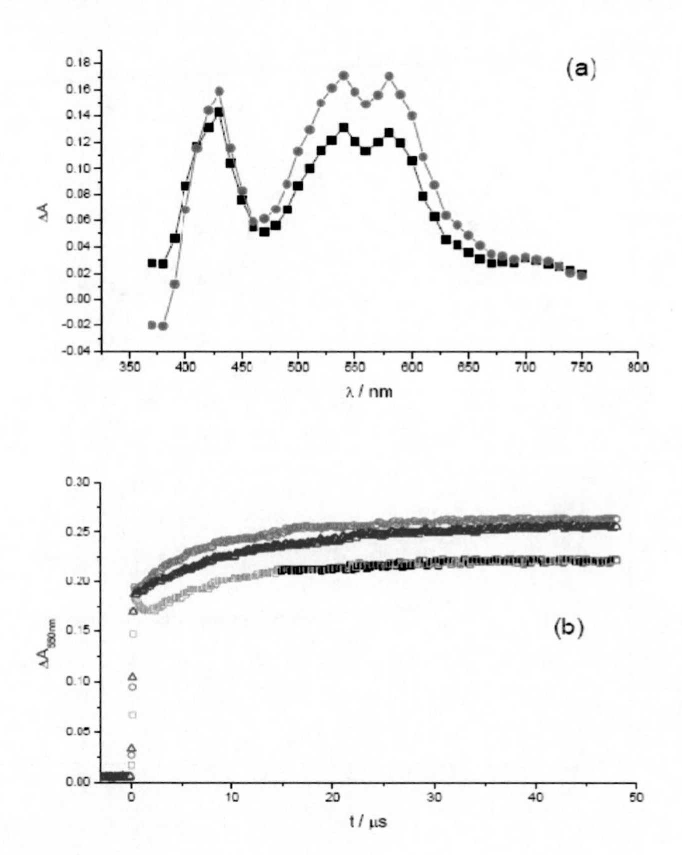

Figure 9. Transient spectra , (a), in the quenching of pyRe$(CO)_3$phen$^+$ MLCT excited state with TEOA 10% (v/v) in CH_3CN. The spectra were respectively recorded with 350 ns (■) and 47 µs (●) delays from the laser irradiation. Traces, (b) at λ_{ob} = 550 nm, were recorded in solutions containing 1(□), 5(○) and 20%(△) (v/v) of TEOA in CH_3CN.

The reductive quenching of the polymer's MLCT excited state by TEOA is represented in eq. 10. Flah photolysis show that an additional formation of Re(I) ligand radical must follow eq. 10. Radicals $TEOA^{\bullet}$ that are formed by the oxidation of TEOA, eq. 11, must react with $-Re(CO)_3phen^+$ chromophores in the polymer, eq. 12, in competition with a fraction of radicals that undergo the reported disproportionation reaction in eq. 13. Flash photolysis experiments show that a fraction of the photogenerated radicals $TEOA^{\bullet}$ also reduce excess $pyRe(CO)_3phen^+$ to $pyRe(CO)_3(phen^{\bullet})$. The disproportionation of the monomer Re(I) ligand radical to form $pyRe(CO)_3phenH_2^+$, eq. 14, occurs with a rate comparable to those recorded for other Re(I) ligand-radical species.

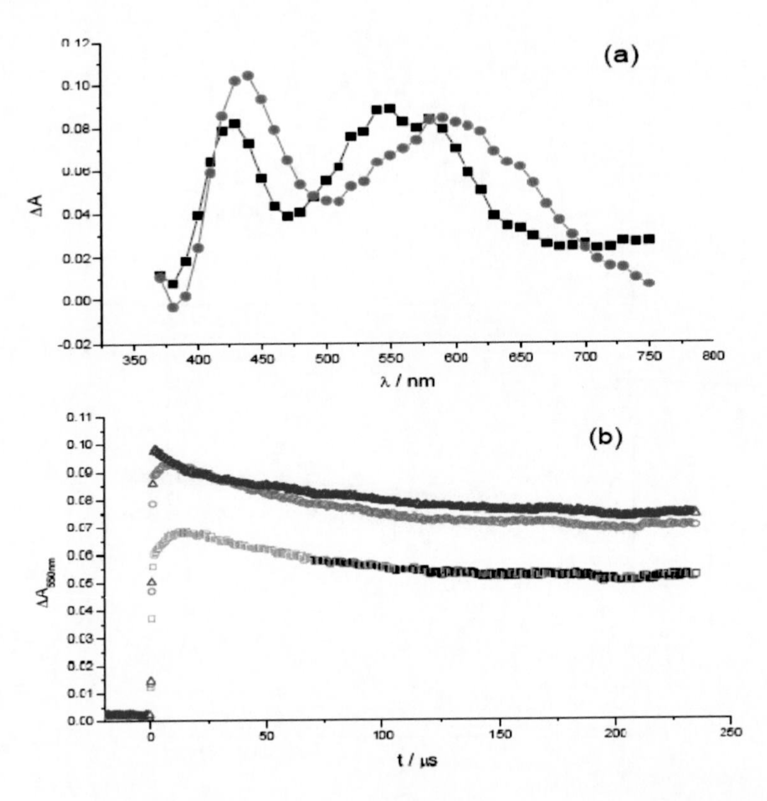

Figure 10. Transient spectra , (a), in the quenching of $\{(vpy)_2\text{-}vpyRe(CO)_3phen^+\}_{n\sim200}$ MLCT excited state with TEOA 10% (v/v) in CH_3CN. The spectra were respectively recorded with 2.5 μs (■) and 231 μs (●) delays from the laser irradiation. Traces, (b) at λ_{ob} = 550 nm, were recorded in solutions containing 1(□), 5(○) and 20%(△) (v/v) of TEOA in CH_3CN.

$$-\left[\underset{py}{\top}\cdots\underset{\underset{Re(CO)_3phen^+}{|}}{\underset{py}{\top}}\cdots\underset{py}{\top}\right]_x-\left[\underset{py}{\top}\cdots\underset{\underset{Re^*(CO)_3phen^+}{|}}{\underset{py}{\top}}\cdots\underset{py}{\top}\right]_y- + y\ TEOA \longrightarrow$$

$$y\ TEOA^{\bullet +} + -\left[\underset{py}{\top}\cdots\underset{\underset{Re(CO)_3phen^+}{|}}{\underset{py}{\top}}\cdots\underset{py}{\top}\right]_x-\left[\underset{py}{\top}\cdots\underset{\underset{Re^I(CO)_3(phen^\bullet)}{|}}{\underset{py}{\top}}\cdots\underset{py}{\top}\right]_y-$$

(10)

$$TEOA^{\bullet +} + TEOA \longrightarrow TEOA^{\bullet} + TEOAH^{+} \quad (11)$$

$$-\left[\underset{py}{\top}\cdots\underset{\underset{Re(CO)_3phen^+}{|}}{\underset{py}{\top}}\cdots\underset{py}{\top}\right]_x-\left[\underset{py}{\top}\cdots\underset{\underset{Re^I(CO)_3(phen^\bullet)}{|}}{\underset{py}{\top}}\cdots\underset{py}{\top}\right]_y- + z\ TEOA^{\bullet}$$

$$\longrightarrow -\left[\underset{py}{\top}\cdots\underset{\underset{Re(CO)_3phen^+}{|}}{\underset{py}{\top}}\cdots\underset{py}{\top}\right]_{x-z}-\left[\underset{py}{\top}\cdots\underset{\underset{Re^I(CO)_3(phen^\bullet)}{|}}{\underset{py}{\top}}\cdots\underset{py}{\top}\right]_{y+z}- + z\ TEOA^{+}$$

(12)

$$2\ TEOA^{\bullet} \longrightarrow TEOA + (OHCH_2CH_2)_2N\text{-}CH{=}CHOH \quad (13)$$

$$2\ pyRe^I(CO)_3(phen)^{\bullet} + 2H^+ \longrightarrow pyRe^I(CO)_3(phenH_2)^+ + pyRe^I(CO)_3(phen)^+ \quad (14)$$

By contrast to the disproportionation of the monomer radical, a partial decay of Re(I) ligand radicals in the polymer to form $-Re^I(CO)_3(phenH_2)^+$ has rate laws in acetonitrile and methanol/water solvents that are respectively of a first and second order on ligand radical concentration [21].

Because the same products are formed in both solvents, it is proposed that the reduction of $-Re^I(CO)_3(phen^\bullet)$ groups proceeds via two sequential processes, one being kinetically of a first order and the other being kinetically of second order. In acetonitrile the former process is the rate determining step whereas in MeOH/H_2O the rate determining step is the latter process. The process that exhibits a first order rate law can be related to the formation of additional vicinal -$Re^I(CO)_3(phen^\bullet)$ groups that disproportionate in a reaction similar to that of the monomer in eq. 14. When formation of vicinal -$Re^I(CO)_3(phen^\bullet)$ groups is fast, electron transfer between such groups is the rate determining step. Because a residual concentration of -$Re^I(CO)_3(phen^\bullet)$ groups survive several minutes, the reactions of $TEOA^\bullet$ leave some of these groups in isolation for either the transfer or the acceptance of electrons. The demise of these residual Re(I) ligand radicals must occur by a slower disproportionation process that possibly demands larger diffusive displacements of polymer strands in a time scale of minutes.

The nature of the products [26] of the photolysis of polymer $\{(vpy)_2\text{-}vpyRe(CO)_3phen^+\}_{n\sim200}$ in the presence of TEOA and/or TEA were investigated by steady-state irradiation at λ_{irr}=350 nm using lamps with $5x10^{-4} \geq I_0 \geq 1x10^{-4}$ Einstein dm^{-3} min^{-1} for periods of photolysis smaller than 100 min in solutions which had been deaerated by bubbling either N_2 or CO_2. Solutions made in DMF developed a blue color during the photolysis due to the appearance of absorptions at $\lambda > 500$ nm (figure 11). When DMF was substituted by CH_3CN as a solvent and the remaining experimental conditions where the same photolysis of the polymer produced dark blue precipitates which could be separated from the reaction mixture and characterized by their UV-vis, IR and elemental analysis. The absence of a ESR signals at 77K demonstrated that the solid products were neither Re(II) nor radical containing products. Several new features were observed in the IR of the final products. The splitting and shift to lower frequencies of the CO stretching was noticed in the solid product formed in DMF solutions deaerated with N_2. Also, the fingerprint of absorptions expected for bridging CO were missing from the IR. New absorptions were observed in the IR spectra of the solid products, one at 1032-1073 cm^{-1} was attributed to groups $(CH_2(OH)CH_2)_2N$- in the structure of the polymer obtained when TEOA was used as the electron donor. An absorption at 1350 cm^{-1} was assigned to the group $-COCH(N(CH_2CH_3)_2)CH_3$ in the product obtained using TEA as electron donor (figure 12). In the elemental analysis of the product, a significant increase of the C and N content, relative to the starting material, is in agreement with the incorporation of such a group to the structure of the polymer. Amounts of CO below the detection limit of the chromatographic procedure were generated when more than 70% of the polymer was converted to the blue product. On the basis of this molar relationship, the quantum yield of CO was negligible by comparison to the quantum yield of Re(I) chromophore converted to product, which was $\phi = (2.0\pm0.2)x10^{-2}$.

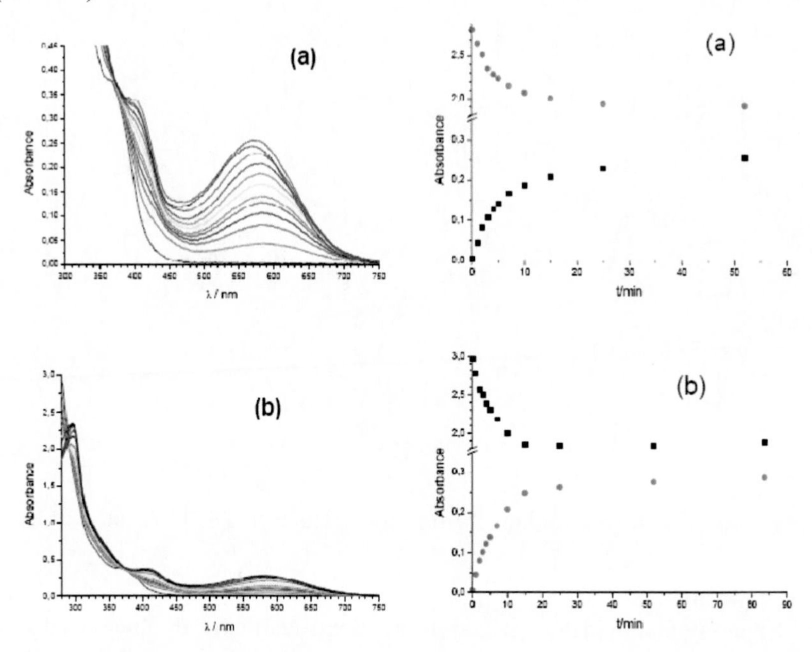

Figure 11. Spectral changes in the UV-vis regions observed during the steady state photolysis of $\{(vpy)_2\text{-}vpyRe(CO)_3phen^+\}_{n\sim200}$ and TEOA, λ_{irr} = 350 nm in CO_2 saturated (a), or N_2 saturated (b), DMF. Left panels show absorbance changes at 278 nm (■) and 580 nm (•) during irradiation time.

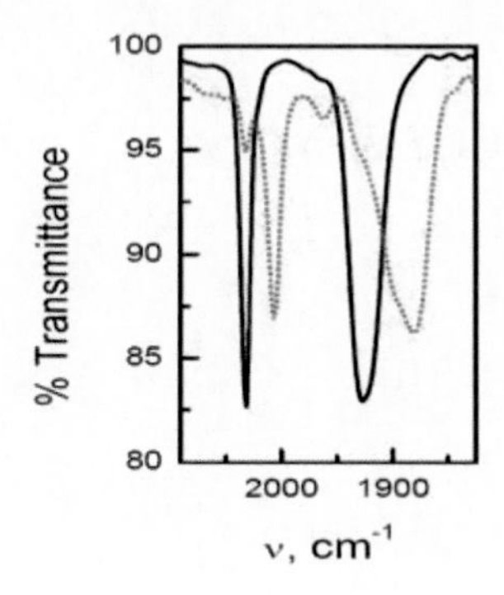

Figure 12. IR spectrum of a N_2 deaerated , DMF solution of $\{(vpy)_2\text{-}vpyRe(CO)_3phen^+\}_{n\sim200}$ in the region of the CO stretching before (—) and after (•••) irradiation at 350 nm.

One structure of the products that satisfies the experimental evidence is that of eq. 15

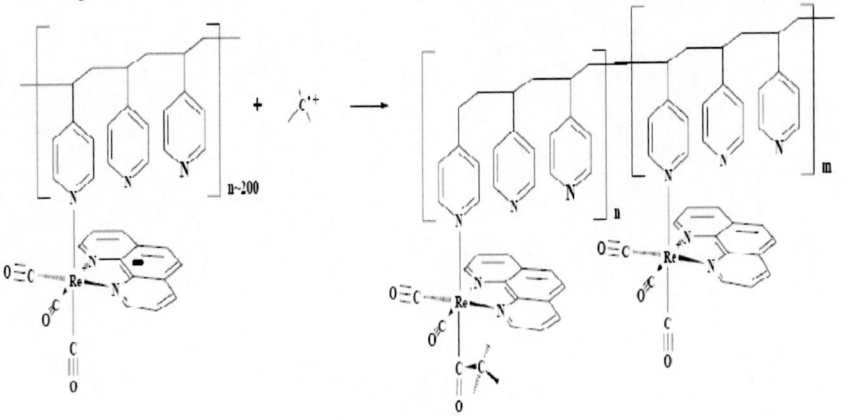

(15)

Where the C-centered radicals from the oxidation of TEA or TEOA are represented by ($C^{\bullet +}$).

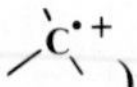

Experimental evidence helped in the identification of the blue products as Re compounds containing alkylmethanal groups in their structure:

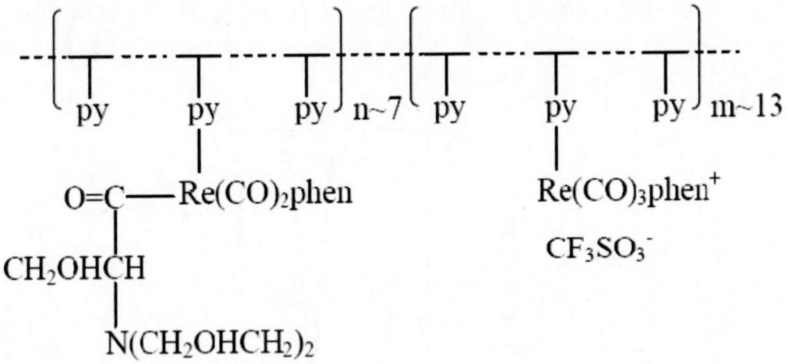

py py py n~7 py py py m~13

$O{=}C$—$Re(CO)_2phen$ $Re(CO)_3phen^+$

CH_2OHCH $CF_3SO_3^-$

$N(CH_2OHCH_2)_2$

Chapter 6

BINDING OF CU(II) ESPECIES TO POLYMER (II)

The addition of a acetonitrile solution of CuX_2 (X= Cl or CF_3SO_3) to an acetonitrile solution of the polymer (II) ([Re(I)] ~ $8x10^{-4}$ M, $[py]_{uncoordinated}$ ~ $1.5x10^{-3}$M) produces a rapid coordination of the Cu(II) species to the uncoordinated pyridines of the Re(I)-$(vpy)_{n\sim600}$ polymer. Indeed, after the mixing of the solutions there is an instantaneous change in the solution color from yellow to green, due to the appearance of a new absorption band. The coordination of the Cu(II) species to the Re(I) polyelectrolyte was followed by the UV-vis spectral changes of the weak ligand field d-d bands of Cu(II) in the range 500-900 nm, where the polymer Re(I)-$(vpy)_{n\sim600}$ has no significant absorption. The addition of $CuCl_2$ to the Re(I) polymer produces the growth of a new absorption band, centered at λ_{max} ~ 725 nm, which is blue shifted respect to the d-d bands of the "free" $CuCl_2$ (λ_{max} ~ 850 nm). After the coordination of Cu(II) is complete, the solution spectrum maximum shifts from λ_{max} ~ 725 nm to λ_{max} ~ 850 nm due to the presence of increasing amounts of uncoordinated $CuCl_2$ in the solution. From a linear fit analysis of the initial and final slopes of the plot A_{725nm} vs. [Cu(II)] it could be calculated that the maximum amount of Cu(II) bound to the polymer was $[Cu(II)]_{b,max}$ = $6x10^{-4}$ M, i.e. between one half and one third of the total initial free pyridine concentration. Similar spectral changes were observed after the addition of a solution of $Cu(CF_3SO_3)_2$ to a solution of the polymer (II) in the same experimental conditions ([Re(I)] ~ $8x10^{-4}$ M, $[py]_{uncoordinated}$ ~ $1.5x10^{-3}$M). The addition of $Cu(CF_3SO_3)_2$ to the Re(I) polymer generates a new absorption band centered at λ_{max} ~ 610 nm which is blue shifted respect to the d-d bands of the "free" $Cu(CF_3SO_3)_2$ ($\lambda_{max} \sim 750$ nm). After the saturation of the polymer with $Cu(CF_3SO_3)_2$, there is a progressive shifting of the absorption maximum from 610 to 750 nm due to the presence of increasing amounts of uncoordinated $Cu(CF_3SO_3)_2$ in the solution. From a linear fit analysis of the initial and final slopes of the plot A_{600nm} vs. [Cu(II)] it could be calculated that the maximum amount of Cu(II) bound to the polymer was $[Cu(II)]_{b,max} = 4x10^{-4}$ M, i.e. less than one third of the total initial free pyridine concentration. It is worthy to note that the coordination of Cu(II) to the polymer Re(I)-$(vpy)_{n\sim600}$ produces an increase in the extinction coefficient of the d-d transitions of the CuX_2 (X= Cl^- or $CF_3SO_3^-$). Quantitative treatment of the absorption data allows the determination of the binding constants between Cu(II) and the pyridines in the Re(I) polyelectrolyte according to different theories [27, 28]. We chose that of McGhee and von Hippel [28], which describes random non-cooperative binding to a lattice:

$$\frac{\nu}{C_f} = K_b \frac{(1-n\nu)^n}{[1-(n-1)\nu]^{n-1}} \tag{16}$$

where ν is the binding ratio C_b/[py], K_b is the binding constant and n is the average size of a binding site (expressed in number of pyridines per Cu(II) specie). In order to use this equation, the concentration of Cu(II) bound (C_b) and Cu(II) free (C_f) for each total concentration of Cu(II) ($C_T = C_b + C_f$) have to be determined. Therefore, C_b was calculated from the UV-vis absorption data as follows: for each total Cu(II) concentration, the difference $A_{polymer+Cu}$ - A_{Cu} must be proportional to C_b and eventually reaches a limiting value for C_T >> C_b. From this we can obtain the ratio $C_b/C_{b,max}$ for each C_T. Finally, with the known values of $C_{b,max}$ (see above) we determined C_b for each value of C_T. From a curve fit analysis of ν/C_f vs. ν (eq. 16) values of $K_b = 2x10^4$ M^{-1} and n = 1.8 and $K_b = 1x10^5$ M^{-1} and n = 3.0 were obtained for the binding of $CuCl_2$ or $Cu(CF_3SO_3)_2$ to the Re(I) polymer respectively (see figure 13).

The coordination of $CuCl_2$ to poly-4-vinylpyridine [29] or to the partially methyl quaternised $(vpy)_{n\sim600}$ [30] has been reported in the literature. The new absorption bands generated after the coordination of CuX_2 to the Re(I)-$(vpy)_{n\sim600}$ polymer have λ_{max} = 725 and 610 nm when X = Cl and CF_3SO_3, respectively. These facts can be compared with the similar shift of the absorption maximum from 770 to 610 nm and an increase of the molar extinction coefficient observed in the stepwise formation of CuL_i^{2+} complexes, (i = 1 to 4) with L = ethylpyridine [30], in water as a solvent. This is because the pyridine produces a stronger ligand field, which causes the absorption band to move from the far red to the middle of the red region of the spectrum. Besides, an increase in the d-d extinction coefficient after coordination of pyridine to Cu(II) should be related to a decrease in the symmetry around the metal center. It is reasonable to assume then that the complexation of $CuCl_2$ to the Re(I)-$(vpy)_{n\sim600}$ polymer to form the polymer Re(I)-$(vpy)_{n\sim600}$ -$CuCl_2$ proceeds with a lower average coordination number than in the complexation of $Cu(CF_3SO_3)_2$ to the Re(I) polymer to form the polymer Re-$(vpy)_{n\sim600}$-$Cu(CF_3SO_3)_2$.

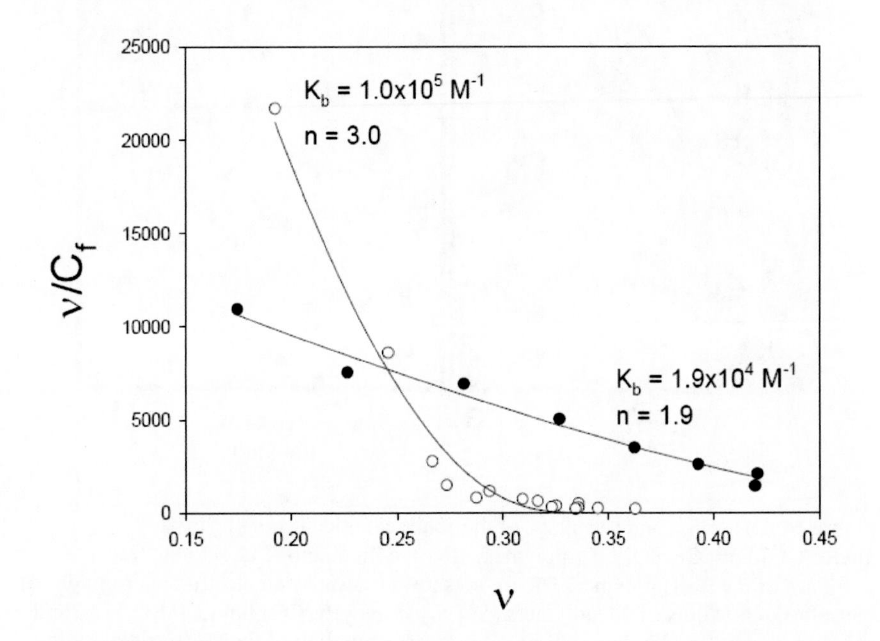

Figure 13. Mc.Ghee- von Hippel plot for the binding between (○) $Cu(CF_3SO_3)_2$ and (●) $CuCl_2$ species and $\{[(vpy)_2\text{-}vpyRe(CO)_3bpy]\ CF_3SO_3\}_{n\sim200}$. The solid lines represent the best fit of the experimental data according to eq. 16.

Indeed, approximate coordination numbers being 2 and 3 were calculated from UV-vis changes for $CuCl_2$ and $Cu(CF_3SO_3)_2$ respectively (see above). Being $CF_3SO_3^-$ a weak ligand it can explain the apparent higher coordination number. Indeed, $CF_3SO_3^-$ is a poor ligand (compared to Cl^-) and it does not compete with pyridine ligands. In fact, in $Cu(CF_3SO_3)_2$ acetonitrile solutions the predominant species should be $Cu(CH_3CN)_4^{+2}$, however, in $CuCl_2$ acetonitrile solutions significant presence of other species should be considered. After the coordination of Cu(II) species to the Re(I) polymer, there is also a change in the morphology of the polymer (figure 14). The polymer Re(I)-$(vpy)_{n\sim600}$ -$CuCl_2$ aggregate to form micelles that are distorted from the spherical shape and whose dimensions are smaller than those micelles formed by the polymer Re(I)-$(vpy)_{n\sim600}$ [20].

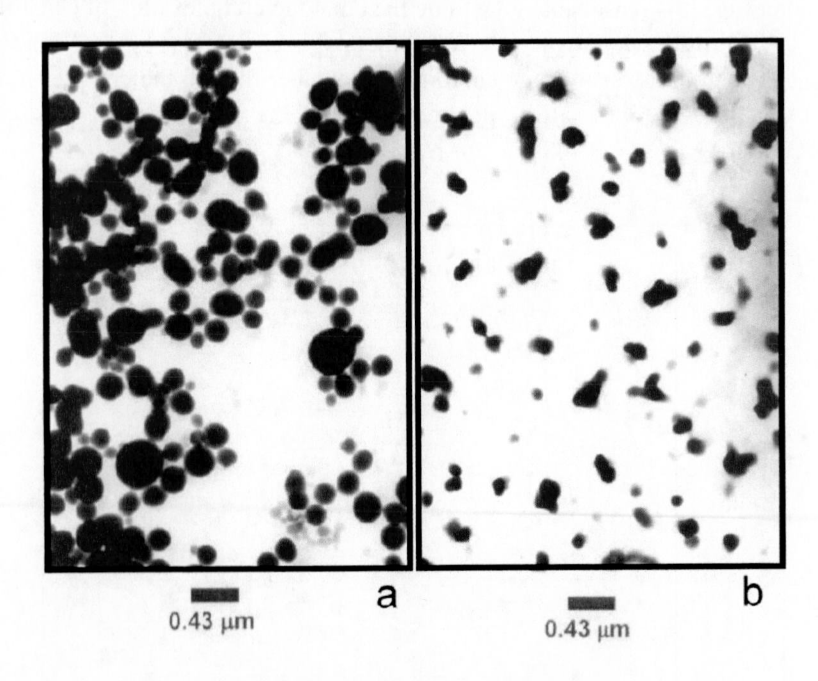

Figure 14. Distortion and shrinkage of the spherical micelles of $\{[(vpy)_2$-vpyRe$(CO)_3$bpy] $CF_3SO_3\}_{n\sim200}$ polymers after the binding of $CuCl_2$ to the free pyridines of the Re(I) polymer. The figure shows transmission electron micrographs of the solvent cast films of the polymers (a) $\{[(vpy)_2$-vpyRe$(CO)_3$bpy] $CF_3SO_3\}_{n\sim200}$ and (b) $\{[(vpy)_2$-vpyRe$(CO)_3$bpy] $CF_3SO_3\}_{n\sim200}$ after saturation of the free pyridines with $CuCl_2$.

Chapter 7

LUMINESCENCE QUENCHING BY CU(II)

The emission spectra of $CF_3SO_3[pyRe(CO)_3bpy]$ and polymer (II) in deoxygenated CH_3CN at room temperature exhibited unstructured bands both centered at 589 nm with identical band shapes. The quenching of the luminescence of the Re(I) polymer by $CuCl_2$ and/or $Cu(CF_3SO_3)_2$ was studied using steady state and time resolved techniques under similar experimental conditions ([Re(I)] ~ $1x10^{-4}$ M) to those utilized in the investigation of the binding of Cu(II) species to the Re(I) polyelectrolyte. As it can be observed from figure 15 the quenching does not follow a Stern-Volmer kinetics. For instance, in the quenching by $Cu(CF_3SO_3)_2$ the ratio Φ_0/Φ follows a sigmoid curve with a limiting value of $(\Phi_0/\Phi)_{max} = 6.3$ at $[Cu(CF_3SO_3)_2]$ ~ $3x10^{-4}$ M, which corresponds to 84% of the total emission quenched. Conversely, Φ_0/Φ in the quenching by $CuCl_2$ deviates upwards respect to the linear Stern-Volmer usual behavior with no limiting value of Φ_0/Φ. Moreover, at $[CuCl_2]$ ~ $3x10^{-4}$ M Φ_0/Φ ~ 40 (corresponding to 97% of the total emission quenched). Figure 16 shows the normalised emission spectra at different [Cu(II)] for the quenching by $CuCl_2$ (figure 16a) and by $Cu(CF_3SO_3)_2$ (figure 16b). As it can be observed from figure 16a, the Re(I)-polymer emission spectra band shape are not altered by $CuCl_2$ addition up to $[CuCl_2] < 6x10^{-5}$M. When the $[CuCl_2] > 8x10^{-5}$M there is a progressive shifting of the emission maximum from 589 to 555 nm. Figure 16b shows that there are hardly any changes in the normalized emission spectra of the Re(I)-polymer after the addition of $Cu(CF_3SO_3)_2$ in the same concentration range as with $CuCl_2$. However, the luminescence quenching of the monomer $CF_3SO_3[pyRe(CO)_3bpy]$ by CuX_2 (X= Cl or CF_3SO_3) follows a typical Stern-Volmer kinetics with $K_{sv,CuCl2} = 4.6x10^3$ M^{-1} and $K_{sv,Cu(CF3SO3)2} = 2.1x10^2$ M^{-1}, respectively (figure 15b). It is noteworthy that no spectral changes occur in the emission spectra of $CF_3SO_3[pyRe(CO)_3bpy]$ in the quenching of its luminescence by CuX_2 (X= Cl or CF_3SO_3).

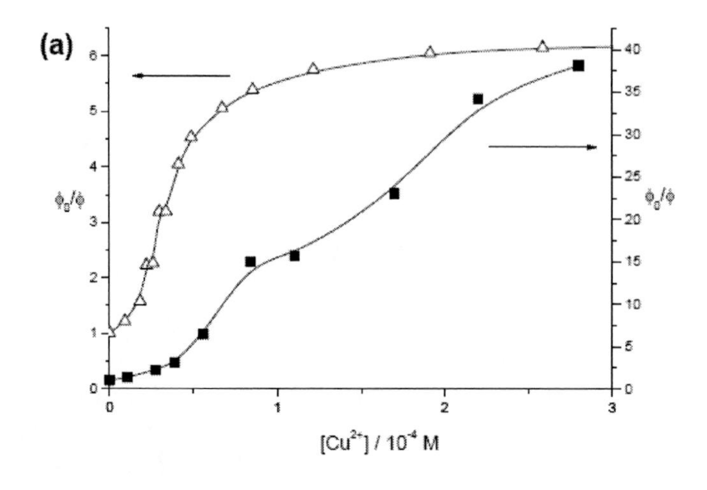

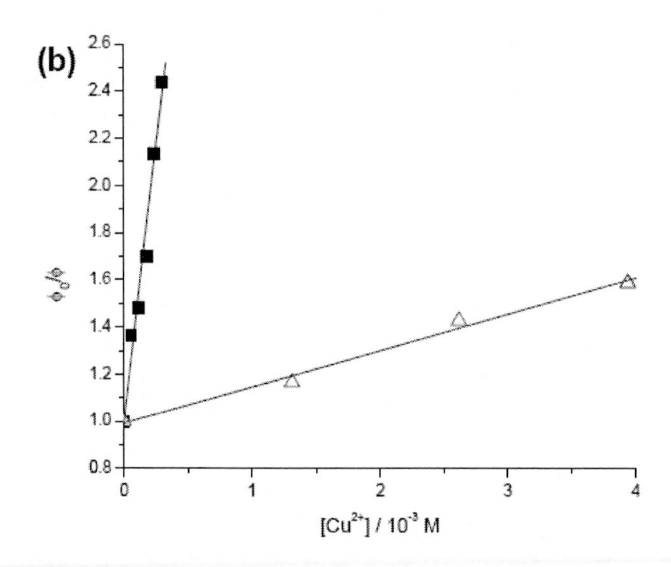

Figure 15. (a) Steady state quenching of $\{[(vpy)_2\text{-}vpyRe(CO)_3bpy]\ CF_3SO_3\}_{200}$'s luminescence by Cu(II) species: (■) Φ_0/Φ (right y-axis), Quencher = $CuCl_2$, (△) Φ_0/Φ (left y-axis), Quencher = $Cu(CF_3SO_3)_2$. (b) Steady state quenching of $pyRe(CO)3(bpy)^+$ by $CuCl_2$ (■) and $Cu(CF_3SO_3)_2$ (△). See text for details.

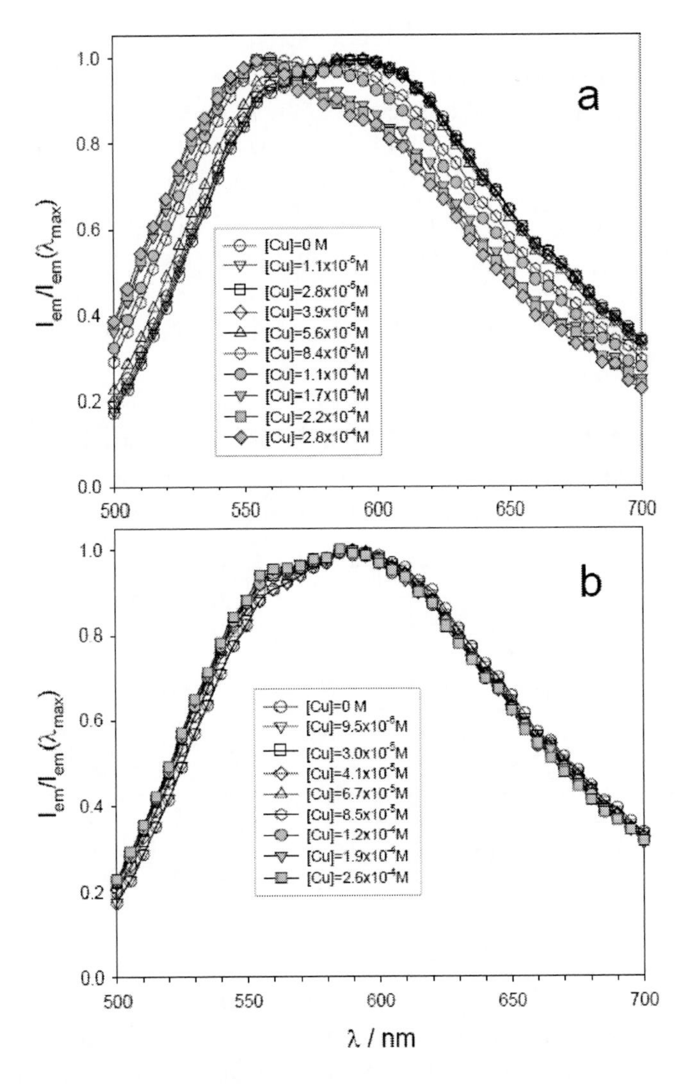

Figure 16. Normalised luminescence spectra of $\{[(vpy)_2\text{-}vpyRe(CO)_3bpy]\ CF_3SO_3\}_{200}$ at different Cu(II) concentrations in the quenching by (a) $CuCl_2$ and (b) $Cu(CF_3SO_3)_2$.

After excitation with a N_2 laser the luminescence decay of the polymer (II) in acetonitrile is monoexponential with a lifetime of τ_{em} = 203 ns [23]. However, after the addition of CuX_2 (X= Cl or CF_3SO_3) the luminescence decay of the Re(I) polymer became biexponential, eq 17:

$$I_{em}(t) = I_1 \times \exp(-t/\tau_1) + I_2 \times \exp(-t/\tau_2) \tag{17}$$

Table 2 shows the values of the preexponential factors (I_1 and I_2) and the lifetimes (τ_1 and τ_2) obtained by a curve fit analysis of I_{em} according to the eq. 17 in the quenching of the luminescence of the Re(I)-polymer by CuX_2 (X= Cl and CF_3SO_3). Inspection to Table 2 shows that as the $Cu(CF_3SO_3)_2$ concentration is increased, the ratio I_1/I_2 decreases. Besides, the longer lifetime τ_1 decreases from 203 ns (in the absence of quencher) to ~130 ns and eventually remains constant. The shorter lifetime also decreases from τ_2 ~ 40 ns to τ_2 ~ 30 ns. At concentrations of $Cu(CF_3SO_3)_2 > 6 \times 10^{-5}$ M both lifetimes τ_1 and τ_2 remain constant. A similar behaviour is followed by the curve fit parameters I_1, I_2, τ_1 and τ_2 in the quenching by $CuCl_2$.

Table 2. Parameters obtained in the curve-fit analysis of the experimental luminescence intensity, $I_{em}(t)$, according to a biexponential decay (eq. 16). See text for details

$[Cu(CF_3SO_3)_2]$/M	I_1/I_2	τ_1/ns	τ_2/ns	$[CuCl_2]$/M	I_1/I_2	τ_1/ns	τ_2/ns
0	--	203[a]	--	0	--	203[a]	--
9.8×10^{-6}	--	195[a]		1.2×10^{-5}	--	166[a]	--
2.0×10^{-5}	1.47	190	44	2.4×10^{-5}	1.44	169	41
3.0×10^{-5}	0.63	148	34	3.6×10^{-5}	0.69	146	33
3.9×10^{-5}	0.39	126	30	4.8×10^{-5}	0.37	122	27
4.9×10^{-5}	0.29	119	29	6.0×10^{-5}	0.20	112	24
5.8×10^{-5}	0.27	114	27				
6.8×10^{-5}	0.25	116	27				
7.8×10^{-5}	0.21	131	27				
8.7×10^{-5}	0.17	131	27				

[a] Monoexponential decays.

The quenching of the Re(I)-$(vpy)_{n\sim600}$ polymer's luminescence by CuX_2 (X= Cl or CF_3SO_3) proceeds via both dynamic and static mechanisms as it can be observed from the comparison of Φ_0/Φ (figure 15a) and τ functional dependences (Table 2) on Cu(II) concentration. For instance, in the luminescence (steady state or time resolved) quenching by $Cu(CF_3SO_3)_2$ we have [Re(I)] ~ 1×10^{-4} M and $[py]_{uncoordinated}$ ~ 2×10^{-4} M and hence $C_{b,max}$ ~ 7×10^{-5} M. From Table 2 it can be observed that at $[Cu(CF_3SO_3)_2]$ ~ 7×10^{-5} M both lifetimes τ_1 and τ_2 achieve constant values. However, figure 15a shows that the quenching measured by steady-state luminescence techniques (increase in the ratio Φ_0/Φ vs. Cu(II) concentration) is still significant at $[Cu(CF_3SO_3)_2] > 7 \times 10^{-5}$ M. A limiting plateau in Φ_0/Φ is achieved only at $[Cu(CF_3SO_3)_2] > 1.5 \times 10^{-4}$ M. Moreover, results from steady state experiments (figure 16a) with $CuCl_2$ must be associated too with more than one quenching process. For instance, when $0 < [CuCl_2] < C_{b,max} \sim 7 \times 10^{-5}$ M the quenching proceeds with no spectral changes in the luminescence spectra. However, when $[CuCl_2] > C_{b,max}$ there is a progressive shifting of the emission maximum to shorter wavelengths. This behaviour could be attributed to some kind of interaction between $CuCl_X^{(2-x)}$ species and the Re(I)-polymer. This effect was not observed in the quenching by $Cu(CF_3SO_3)_2$ (compare figures 16a and 16b). On the other hand, the monomer $CF_3SO_3[pyRe(CO)_3bpy]$ luminescence is quenched by $CuCl_2$ and by $Cu(CF_3SO_3)_2$ following typical Stern-Volmer kinetics with no spectral changes in the luminescence spectrum even at higher [Cu(II)] than those used when quenching the luminescence of the Re(I)-$(vpy)_{n\sim600}$ polymer.

The luminescence quenching by $Cu(CF_3SO_3)_2$ in the Re(I) polymer is far more efficient than in the monomer. There are two different reasons that can explain such a difference. First, there should be an enhancement of the energy transfer rate constant due to a close vicinity between the Re(I) chromophore and the Cu(II) bound to a near pyridine in the polymer. Second, an enhancement of the spectral overlap integral (J) between the emission spectrum of the Re(I) chromophore and the new absorption band which appears after the coordination of the Cu(II) ion to the Re(I)-$(vpy)_{n\sim600}$ polymer (note that the Dexter's exchange mechanism for energy transfer predicts that $k_{et} \alpha \exp(-R)$ while the Forster's dipole-induced energy transfer theory predicts that $k_{et} \alpha R^{-6}$, i.e. both theories predict an increase of k_{et} as R decreases. Besides, in both Dexter's and Forter's theories $k_{et} \alpha J$) [17]. The quenching of the monomer $CF_3SO_3[pyRe(CO)_3bpy]$ luminescence by $CuCl_2$ follows a typical Stern-Volmer kinetics and taking into account its luminescence lifetime [23] a bimolecular rate constant $k_{q,CuCl2} = 2 \times 10^{10} M^{-1} s^{-1}$ can be calculated, the value of which is close to the diffusional limit in acetonitrile [31]. A far lower bimolecular rate constant $k_{q,Cu(CF3SO3)2} = 8 \times 10^{8} M^{-1} s^{-1}$ can be calculated from the quenching by $Cu(CF_3SO_3)_2$ (figure 15b). The poor spectral overlap integral (J) between absorption spectrum of $Cu(CF_3SO_3)_2$ and $CF_3SO_3[pyRe(CO)_3bpy]$ emission spectrum should be responsible for this distinct photophysical behavior [19].

The binding of Cu(II) to the Re(I) polymer produces the breakage of the micelles and yields a wide distribution of particle sizes ranging from a few nanometers at polymer scales to few hundreds of nanometers covering micelle scales. In fitting the luminescence decay by a biexponential function the parameters τ_1 and τ_2 may then be viewed as "mean" lifetimes representing average contributions from the different species [19].

Chapter 8

THEORETICAL OUTLINE OF FLUORESCENCE RESONANCE ENERGY TRANSFER (FRET) IN POLYMERS

Resonance energy transfer (RET) is a widely prevalent photophysical process through which an electronically excited 'donor' molecule transfers its excitation energy to an 'acceptor' molecule such that the excited state lifetime of the donor decreases. If the donor happens to be a fluorescent molecule RET is referred to as fluorescence resonance energy transfer, FRET. In FRET, the energy may be passed nonradiatively between molecules over long distances (10 – 100 Å). The acceptor however may or may not be fluorescent.

Resonance energy transfer is a non-radiative quantum mechanical process and requires fluorescence emission spectrum of the donor molecule (D) to overlap with the absorption spectrum of the acceptor (A), and the two to be within the minimal spatial range for the donor to transfer its excitation energy to the acceptor. The Förster theory is based on the equilibrium Fermi-golden rule approach, where the transfer of excitation energy is regarded to be the transition between the electronic states $\phi^{D}_{e}\,\phi^{A}_{g}$ and $\phi^{D}_{g}\,\phi^{A}_{e}$ (where '*g*' and '*e*' stand for ground and excited state respectively) promoted via a coulombic interaction, a long-range dipole–dipole intermolecular coupling between D and A. The key assumptions of Förster formulation are: (a) A dipole–dipole approximation can be employed for electronic coupling between D and A; (b) vibrational relaxation after electronic excitation of donor takes place on a much faster time-scale as compared to RET; (c) coupling of molecules to the surroundings is much stronger than coupling between D and A, ensuring that FRET is an irreversible and incoherent process.

The rate constant for transfer between a donor and an acceptor at a distance R is defined as [32]:

$$k_{ET} = \frac{1}{\tau_D}\left(\frac{R_F}{R}\right)^6 \tag{18}$$

where τ_D is the excited state lifetime of the donor in the absence of transfer and R_F is the Förster critical radius, defined to be the distance at which the efficiency of energy transfer from D to A becomes 50%. R_F is given by the spectral overlap between the fluorescence spectrum of the donor and the absorption spectrum of the acceptor and can be calculated according to eq. 19 [32]:

$$\left(\frac{R_F}{cm}\right)^6 = 8.79 \text{ x } 10^{-25}\ \frac{\kappa^2\phi_D}{n^4}\ \int\frac{f_D(\upsilon)\ \varepsilon_A(\upsilon)\ d\upsilon}{dm^3 mol^{-1} cm^3\ \upsilon^4} \tag{19}$$

Where κ^2 is an orientation factor equal to 2/3 for an isotropic angular distribution, n is de refractive index of the medium, ϕ_D is the quantum yield of the emission donor, $\varepsilon_A(\upsilon)$ is the decadic extinction coefficient of the acceptor at wavenumber υ, and $f_D(\upsilon)$ is the emission spectrum of the donor such that $\int f_D(\upsilon)\ d\upsilon$ is unity.

So far, this is the expression for the energy transfer constant (k_{ET}) in the current Förster´s energy transfer theory for a definite distance R between D and A. The problem in evaluating k_{ET} from luminescence quenching data of the donor is much more complicated in a polymer where D and A are distributed at random in the polymer strand.

However, the time dependence of D and A fluorescence after pulsed excitation can still be related to the interchromophore distance via the rate of the Förster´s energy transfer theory.

When fluorophore molecules are immobilized in polymer backbones, the fluorescence decay is influenced by interactions between fluorophores and the polymer matrix. As most polymers are nonuniform media , i.e. there is no long-range order, the fluorescence decay profiles of the molecules can be assumed to be an average over a distribution of relaxation rates. A consequence of this is that the decay function of the ensamble of molecules becomes nonexponential. The time dependent probability P(t) for an excited luminophore to be in an excited state at a time t after excitation is the solution to the differential eq. 20:

$$-\frac{dP(t)}{dt} = P(t)(k_r + k_{nr} + k_{ET}) \tag{20}$$

Where k_r is the radiative rate constant, k_{nr} is the rate for the nonradiative relaxation of the luminophore and k_{ET} is the quenching rate constant due to energy transfer. In evaluating k_{ET}, it has to be assumed that the quenching of the donor D is brought about by interactions with neighboring regions of the polymer containing the acceptor molecules A. The influence of these interactions depends on the distance between D and A. Then the quenching rate for the ith donor D_i is the sum over the distance dependent interactions with j quenching sites :

$$k_{ET}(D_i) = \frac{1}{\tau_D}\sum_j (R_F / R(D_i, j))^6 \tag{21}$$

When substituting k_{ET} into eq. 20 the resulting differencial equation can be solved if the distribution of distances is assumed to be isotropic and homogeneous [33]. In those mathematical treatments of the problem, the polymer concentration is considered low enough to neglect interactions between the chromophores attached to adjacent polymer chains. As the distance R between D and A is dependent on the chain conformation, an average is performed over the conformations with the use of a $g_N(R)$ function, which is the probability density of finding a polymer containing N units in which the distance between the chain ends is R.

The solution to the differential equation, corresponding to a dipole-dipole interaction in three dimensions gives [33]:

$$N_t = N_0 \exp\left[-\frac{t}{\tau_D} - a\sqrt{t/\tau_D}\right] \tag{22}$$

Here N_t is the number of molecules which survived excitation at time t and a is a parameter proportional to the density of acceptor quenching sites which can be calculated according to [33] eq. 23:

$$a = \frac{4}{3}\pi^{3/2}\rho R_F^3 \tag{23}$$

Where ρ stands for the number of quenching sites per volume. Eq. 22 relates the form of the decay curve to a certain quenching mechanism (i.e. Forster´s dd energy transfer) and two structural quantities, namely the density of quenching sites and the critical radius R_F. The number of quenching sites within the critical radius is given by [33], eq. 24

$$N = \frac{a}{\sqrt{\pi}} \tag{24}$$

Finally, the efficiency of energy transfer can be calculated with the aid of eq. 25 according to:

$$E_T = 1 - \frac{\phi_D}{\phi_D^0} \tag{25}$$

Chapter 9

RESONANCE ENERGY TRANSFER IN THE SOLUTION PHASE PHOTOPHYSICS OF POLYMERS (VII)-(IX)

The chemical structures of polymers **(V)-(IX)** are shown in figure 17. Deaerated solutions of the polymers **(V)-(IX)** with total concentration of the Re(I) chromophores equal to or less than 1 x 10^{-4} M in CH_3CN were irradiated at 380 nm to record the emission spectrum. The emission spectrum of polymer **(V)** in deoxygenated CH_3CN at room temperature exhibited an unstructured band centered at 520 nm. Polymer **(VI)** is non-luminescent. Polymers **(VII)-(IX)** have luminescence spectra, which are the same in shape to that of polymer **(V)**. The luminescence quantum yield (ϕ_{em}) of polymer **(V)** is around 0.03. Polymer **(VII)** has a ϕ_{em} that is nearly $^1/_3$ lower than that of polymer **(V)**. ϕ_{em} decreases monotonically from polymer **(VII)** to polymer **(IX)**. Table 3 summarizes all the measured ϕ_{em}. It shows also the ϕ_{em} measured with molar mixtures of polymers **(V)** and **(VI)**, i.e. to obtain the same p/(n+p) in the mixture as there is in polymers **(VII)-(IX)**. ϕ_{em} determined for the mixtures 90%-**(V)** + 10%-**(VI)**, 75%-**(V)** + 25%-**(VI)** and 50%-**(V)** + 50%-**(VI)** are noticeably higher than those of polymers **(VII), (VIII)** and **(IX)**, respectively. For instance, while ϕ_{em} of the mixture 50%-**(V)** + 50%-**(VI)** is nearly ½ ϕ_{em} of polymer **(V)**, ϕ_{em} of polymer **(IX)** is nearly two orders of magnitude lower than that of polymer **(V)**.

Table 3. Photophysical properties of polymers (V), (VI), (VII), (VIII) and (IX) in acetonitrile at room temperature

	ϕ_D[a, b]	τ_{fast}, ns[c]	τ_{slow}, μs[c]	Energy transfer efficiency $E_T^j = 1 - \frac{\phi_D^j}{\phi_{mb}^j}$
Polymers				
(V)	0.035	$(7.4 \pm 0.9)x10^2$	3.4 ± 0.5	
(VI)	$< 10^{-4}$	0.23 ± 0.01[d]	--	
(VII)	0.013	$(1.1 \pm 0.2)x10^2$	1.13 ± 0.07	0.58
(VIII)	0.0038	50 ± 10	0.59 ± 0.07	0.85
(IX)	0.0007	< 10	0.047 ± 0.006	0.94
Molar blends				
90%-(V) + 10%-(VI)	0.031			
75%-(V) + 25%-(VI)	0.026			
50%-(V) + 50%-(VI)	0.0125			

[a] Emission quantum yields of polymers **(V), (VI), (VII), (VIII)** and **(IX)**. Error ± 10%.

[b] Emission quantum yields, ϕ_D, measured with molar blends of polymers **(V)** and **(VI)** in order to obtain the same p/(n+p) in the blend as there is in polymers **(VII)-(IX)**. Error ± 10%. See text for details.

[c] Obtained from a curve fit analysis with two exponentials from transient absorbance decays in flash photolysis experiements (λ_{ex}= 351 nm).

[d] Obtained from a monoexponential decay of the transient absorbance in femtosecond laser photolysis experiments (λ_{ex}= 387 nm).

Transient absorption spectra in the 15 ns to microsecond time domain were recorded with a 351 nm excimer laser flash photolysis set-up. This excitation wavelength produces optical excitation of the MLCT absorption bands of the polymers. When N_2-deaerated acetonitrile solutions of the polymers **(V)**, **(VII)**, **(VIII)** and **(IX)** were irradiated at 351 nm, the transient spectra observed after the 10 ns irradiation decayed biexponentially over a period of several microseconds. The oscillographic traces were fitted to two exponentials with lifetimes τ_{fast} and $\tau_{slow.}$ The lifetimes τ_{fast} and τ_{slow} are collected in Table 3. Transient spectra recorded with either polymer at times immediately after the laser pulse decay, i.e. 20 ns (showing $\Delta A_{t=0}$ values vs. wavelength), showed the spectra of the $^3MLCT_{Re \rightarrow tmphen}$ excited states decaying by radiative and non-radiative processes. The transient spectra generated when these polymers were irradiated at 351nm under the same photochemical conditions (i.e. [Re(I)] and laser energy/pulse) are shown in figure 18. It can be observed that transients generated after 351 nm excitation of polymers **(V), (VII), (VIII)** and **(IX)** have the same spectral features albeit the initial amount (proportional to $\Delta A_{t=0}$) decreases from **(V)** to **(IX)**. Luminescence lifetimes of polymers **(V), (VII), (VIII)** and **(IX)** were measured in their CH_3CN deoxygenated solutions using a flashfluorescence equipment with λ_{exc} = 337 nm. The decay of the luminescent profiles for polymer **(V)** were monoexponential with a lifetime of τ_{em} = 5.12 μs. However, the decay of the luminescent profiles for polymers **(VII)-(IX)** were nonexponential and were fitted with eq. 22.

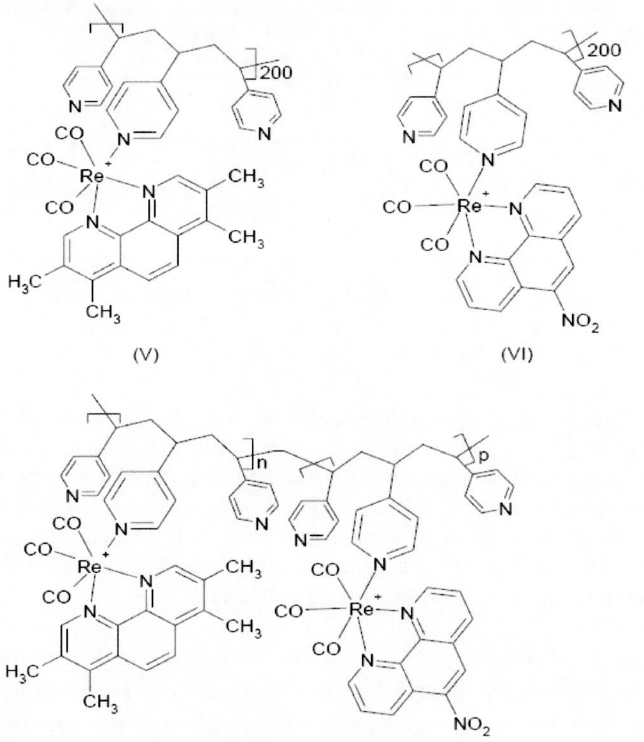

(V)

(VI)

(VII)-(IX)

Figure 17. Chemical structures of polymers . (V): $[(vpy)_2\text{-}vpyRe(CO)_3(tmphen)]_{n\sim200}(CF_3SO_3)_{n\sim200}$; (VI): $[(vpy)_2\text{-}vpyRe(CO)_3(NO_2\text{-}phen)]_{n\sim200}(CF_3SO_3)_{n\sim200}$; (VII): $[(vpy)_{m\sim400}\text{-}(vpy\text{-}Re(CO)_3(tmphen))_{n\sim180}(vpy\text{-}Re(CO)_3(NO_2\text{-}phen)_{p\sim20}](CF_3SO_3)_{n+p\sim200}$; (VIII): $[(vpy)_{m\sim400}\text{-}(vpy\text{-}Re(CO)_3(tmphen))_{n\sim150}(vpy\text{-}Re(CO)_3(NO_2\text{-}phen)_{p\sim50}](CF_3SO_3)_{n+p\sim200}$; (IX): $[(vpy)_{m\sim400}\text{-}(vpy\text{-}Re(CO)_3(tmphen))_{n\sim100}(vpy\text{-}Re(CO)_3(NO_2\text{-}phen)_{p\sim100}](CF_3SO_3)_{n+p\sim200}$.

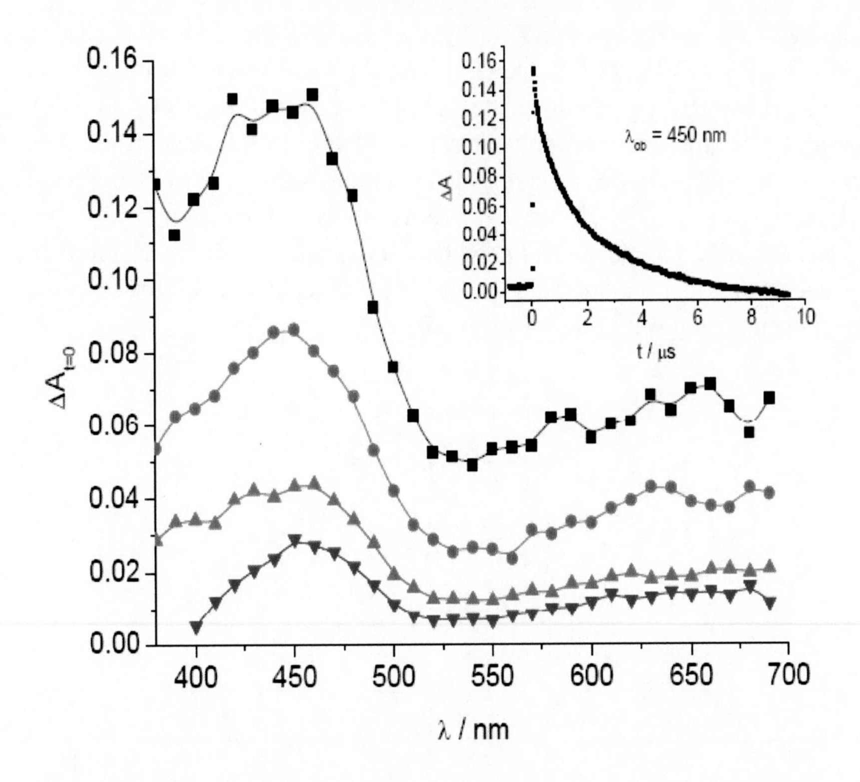

Figure 18. Difference absorption spectrum of the ^{3}MLCT excited states of the -$Re(CO)_3(tmphen)^+$ pendants in polymers (V) (■) , (VII) (●), (VII) (▲) and (IX) (▼). Transient spectra recorded with either polymer at times immediately after the laser pulse decay showing $\Delta A_{t=0}$ vs. λ . The solutions of the polymers in CH_3CN contained a concentration of Re(I) chromophores of [Re(I)] = 5.0 x 10^{-4} M, and it was flash irradiated at 351 nm. The inset is a typical oscillographic trace revealing the biexponential decay of the absorbance at λ_{ob} = 450 nm.

Transient absorption spectra in the femtosecond to nanosecond time domain were recorded with a Ti:sapphire laser flash photolysis instrument providing 387 nm laser pulses for the irradiation of the polymer. The femtosecond to nanosecond transient spectra of the excited states produced ~4 ps after the 387 nm flash irradiation of 1.5 x 10^{-6} M of polymer **(VI)** (i.e. [Re(I)] = 3 x 10^{-4}M) are shown in figure 19. The transient spectrum consists of three absorption bands, two with maxima at 450 and 600 nm, respectively and a third one with λ_{max} > 750 nm. It decays monoexponentially over a period of 1600 picoseconds with a lifetime of 230 ps.

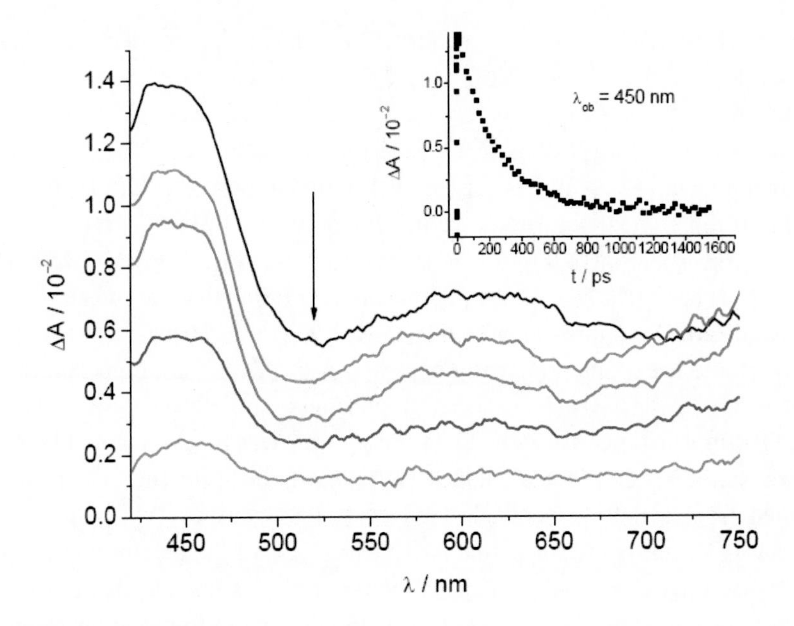

Figure 19. Time resolved absorption spectra recorded on a femtosecond to nanosecond time scale after the 387 nm flash irradiation of polymer (VI) solutions in CH_3CN. The spectra were recorded at 2, 60, 100, 200 and 400 ps time delays after the laser pulse. The solutions of the polymers in CH_3CN contained a concentration of Re(I) chromophores of [Re(I)] ~ 3.0 x 10^{-4} M. The spectra evolved in the direction of the arrow. The inset is a typical oscillographic trace revealing the monoexponential decay of the absorbance at λ_{ob} = 450 nm.

The photophysical properties of polymers **(V)** and **(VI)** are quite different. Polymer **(V)** has a luminescence quantum yield of ϕ_{em} ~ 0.03 and a luminescence lifetime of ~ 5μs, a value that is between ½ and $^1/_3$ of that of the corresponding monomer [34]. Though a longer lifetime for the monomer than for the polymer could be associated with the availability of new deactivation pathways for the MLCT in the polymer due to vibration modes present in the poly(4-vinylpyridine) backbone [23].

On the other hand, polymer **(VI)** is non-luminescent (ϕ_{em} < $1x10^{-4}$). Moreover, after excitation of a polymer **(VI)** solution in CH_3CN with λ_{exc} = 387 nm, the transient generated decayed very fast with a lifetime of 230 ps. Taking into account the spectral similarities between the spectrum of the $MLCT_{Re\rightarrow NO2\text{-}phen}$ in the monomers $LRe(CO)_3(NO_2\text{-phen})^+$ [35] with that of figure 18, the transient generated in the femtosecond to nanosecond time domain of polymer **(VI)** can be assigned to the $^3MLCT_{Re\rightarrow NO2\text{-}phen}$ excited state of $Re(CO)_3(NO_2\text{-phen})^+$ pendants in the polymer decaying non-radiatively. Moreover, the $^3MLCT_{Re\rightarrow NO2\text{-}phen}$ excited states are very short-lived: 7.6, 170, and 43 ps for L = Cl^-, 4-Etpy, and imH, respectively, in CH_3CN solutions [35]. In this regard, the lifetime of the $MLCT_{Re\rightarrow NO2\text{-}phen}$ in the complex with L= 4-Etpy is very similar to that of polymer **(VI)**.

From the comparison of ϕ_{em} values for polymers **(VII), (VIII)** and **(IX)** with the corresponding ϕ_{em} of the molar blends 90%-**(V)** + 10%-**(VI)**, 75%-**(V)** + 25%-**(VI)** and 50%-**(V)** + 50%-**(VI)** it can be inferred that the substitution of pendants $Re(CO)_3(tmphen)^+$ by pendants $Re(CO)_3(NO_2\text{-phen})^+$ in the poly-4-vynilpyridine backbone produces a decrease in ϕ_{em} that is in proportion to the number of $Re(CO)_3(NO_2\text{-phen})^+$ pendants relative to that of $Re(CO)_3(tmphen)^+$ pendants due to an energy transfer process that involves the excited states $MLCT_{Re\rightarrow tmphen}$ and $MLCT_{Re\rightarrow NO2\text{-}phen.}$ This energy transfer is justified from a thermochemical stand point because the UV-vis spectra of polymers **(V)** and **(VI)** show that $^1MLCT_{Re\rightarrow tmphen}$ is higher in energy than $^1MLCT_{Re\rightarrow NO2\text{-}phen}$ excited state. This higher energy is not surprising because the NO_2^- group, which is withdrawing electronic charge from the phen ligand makes the charge transfer from the Re^I to the NO_2-phen energetically more favoured than that of Re^I to phen. The situation is reversed with the tmphen ligand as in the chromophore $Re(CO)_3(tmphen)^+$ the methyl groups are injecting electronic charge to the substituted phen ligand and the charge transfer from the Re^I to tmphen is energetically less favoured than that of Re^I to phen. On the other hand, there is no experimental evidence of a charge transfer in the quenching of the $^3MLCT_{Re\rightarrow tmphen}$ as no photoproducts and/or intermediates (i.e. the formation of $Re(CO)_3(tmphen)^{+2}$ and $Re(CO)_3(NO_2\text{-phen})^\bullet$ species) were observed after the decay of $^3MLCT_{Re\rightarrow tmphen}$ in flash photolysis experiments. The quenching of the $^3MLCT_{Re\rightarrow tmphen}$ by the $Re(CO)_3(NO_2\text{-phen})^+$ pendants must be an energy transfer process rather than an electron transfer one.

ϕ_{em} of the molar blends 90%-**(V)** + 10%-**(VI)**, 75%-**(V)** + 25%-**(VI)** and 50%-**(V)** + 50%-**(VI)** can be compared to the ϕ_{em} values that would have been expected for polymers **(VII), (VIII)** and **(IX)** if no energy transfer between $[Re(CO)_3(tmphen)^+]^*$ and $Re(CO)_3(NO_2\text{-phen})^+$ pendants had occurred, eqs. 26-27:

$$\phi_D = \left(\frac{A}{A_T}\right)\phi_D^{\ 0} \tag{26}$$

$$\frac{A}{A_T} = \frac{n\varepsilon_{\lambda,n}}{n\varepsilon_{\lambda,n} + m\varepsilon_{\lambda,m}} \tag{27}$$

Where A is the absorbance of the donor $Re(CO)_3(tmphen)^+$ at wavelength λ, A_T is the total absorbance of the solution and $\varepsilon_{\lambda,n}$ and $\varepsilon_{\lambda,m}$ are the molar extinction coeffients of the donor $Re(CO)_3(tmphen)^+$ and the acceptor $Re(CO)_3(NO_2\text{-phen})^+$ at wavelength λ, respectively, and ϕ_D^0 is the emission quantum yield of the donor, i.e. 0.035 for polymer **(V)**. Given the values of $5.5x10^5$ and $7.8x10^5$ $M^{-1}cm^{-1}$ for the extinction coefficients of polymers **(V)** and **(VI)** at 380 nm, ϕ_D values of 0.0298, 0.0236 and 0.0142 can be calculated for polymers **(VII), (VIII)** and **(IX)**, respectively. Those ϕ_D values are similar to the experimental values measured for the blends 90%-**(V)** + 10%-**(VI)**, 75%-**(V)** + 25%-**(VI)** and 50%-**(V)** + 50%-**(VI)**, which are 0.031, 0.026 and 0.0125, respectively (see Table 3). Therefore, it can be concluded that there is not bimolecular (i.e. intermolecular) quenching between excited $Re(CO)_3(tmphen)^+$chromophores in polymers **(V)** and acceptors $Re(CO)_3(NO_2\text{-phen})^+$ in polymers **(VI)** in the blends.

Triplet-triplet (T-T) RET is overall a spin-allowed process; spin is conserved between the initial state $^3(^3D^*\text{-}^1A)$ and the final state $^3(^1D\text{-}^3A^*)$, where D stands for the donor and A for the acceptor of the energy transfer process. It is known that T-T RET cannot be mediated by the Coulombic interaction because it is spin-forbidden. Therefore, spin selection rules for dipole-dipole transfer are that no change of spin for either donor or acceptor transitions can occur [36]. A Coulombic interaction can, in principle, promote T-T RET via spin-orbit coupling terms in the Hamiltonian. However, spin-orbit coupling–mediated Coulombic coupling would usually be much smaller than the normal Coulombic interaction between the singlet states of donor and acceptor [36].

On the other hand, spin forbidden transitions in the donor have been observed leading to an enhanced lifetime and long range transfer while no such compensation applies when the transition is spin forbidden in the acceptor [32]. FRET studies in organic polymers are usually discussed in terms of dipole-dipole interactions since D and A are pendants attached to the organic backbone, no diffusion and encounter complexes are likely to occur between them and Dexter´s exchange [32] mechanism for energy transfer is disfavored against the dipole-dipole interaction. When the molecular species are transition metal complexes in fluid solutions and when metal centered ligand field (or d-d) excited states are involved, the basic principles for description of the energy transfer reaction coordinate are not altogether clear. Thus the much studied and well understood dipole-dipole, or Förster mechanism for energy transfer is not strictly applicable when spin forbidden excited states are involved. However, FRET has also been employed in energy transfer studies involving inorganic systems where triplet (or higher spin multiplicity, as in the case of Lanthanides) states may be involved in the energy transfer process [33, 37-49].

Polymers **(VII), (VIII)** and **(IX)** consist of donors $Re(CO)_3(tmphen)^+$ and acceptors $Re(CO)_3(NO_2\text{-phen})^+$ distributed at random through coordination to the pyridines of the $(vpy)_{n\sim600}$ polymer backbone. Locally, there should be regions with no pyridine spacer (D-A , D-D, A-A), a single pyridine spacer (D-py-A , D-py-D, A-py-A) and two or more spacers (D-py-py-A, D-py-py-D, A-py-py-A), etc. MLCT lifetimes and energy transfer dynamics are dependent on the local environment, and there is a distribution of sites on individual polymer strands. According to a molecular modeling calculation [50] the average periphery-to-periphery distance between nearest neighbors might be about 7 Å in these polymers. Moreover, local segmental motions could decrease that distance to well below 7 Å. Therefore, at distances between nearest neighbors of 7 Å or lower, the detailed energy-transfer mechanism, whether Förster (coulomb) or Dexter (exchange) or a combination of the two is unknown. Given its $1/R^6$ distance dependence, Förster transfer is favored over Dexter transfer at long distances since Dexter transfer varies exponentially with distance. As Förster transfer requires that spin be conserved separately at the energy transfer donor and acceptor, this condition could be overcome if spin-orbit coupling is of importance, as for instance in Ru(II) and Os(II) polystyrene polymers [51]

To elucidate further the nature of the energy transfer process in our polymeric systems, we performed a quenching experiment of polymer **(V)** luminescence using $ClRe(CO)_3(NO_2\text{-phen})$ as energy acceptor in a solvent mixture (Glycerol/Ethanol/Dichloromethane, 5:3:2 v:v:v, hereafter referred as GED) where diffusion was somewhat restricted. For that solvent, with a viscosity of η=0.23 poise, a diffusion rate constant $k_{diff} = 2.0\text{x}10^8\ M^{-1}s^{-1}$can be calculated [17]. This value of k_{diff} is similar to that of 1,2 ethanediol (η=0.20 poise) [17], $k_{diff} = 3.0\text{x}10^8\ M^{-1}s^{-1}$. As in GED solvent system quenching is observed at $ClRe(CO)_3(NO_2\text{-phen})$ concentrations as low as $2\text{x}10^{-4}$ M [50], a diffusion lifetime $\tau_{diff} = 1/(\ k_{diff}[Q]) = 25\ \mu s$ can be calculated for that quencher concentration. This τ_{diff} is nearly 5 times longer than the luminescence lifetime of polymer **(V)**. At a quencher concentration ~$2\text{x}10^{-4}$ M the mean distance between molecules is ~110 Å (eq. 1) Therefore, within the excited state lifetime of the polymer, quencher molecules will be able to diffuse only ~20-30 Å and on the average they will be far apart (distances > 70 Å) from the luminescent center. At those large separation distances between D and A, Dexter's mechanism of energy transfer is disfavored against Förster's.

Luminescence decay profiles of polymers **(VII), (VIII)** and **(IX)** were fitted using eq. 22 giving the values of a = 1.90, 4.6 and 19.3, respectively with τ_D = 5.12 μs (Table 3). Values of N ~ 1.1; 2.6 and 11 can be calculated for polymers **(V)-(VII)** using the vales of a = 1.90, 4.6 and 19.3 respectively.

A calculation of R_F for the present system gives a value of 10.7 Å. This value is in the lower limit of R_F values (typically between 10 and 70 Å) due to the poor overlap between the emission spectrum of the donor $^3MLCT_{Re\rightarrow tmphen}$ and the absorption spectrum of the acceptor $Re(CO)_3(NO_2\text{-phen})^+$. Using this value of R_F and the values of a obtained above from the luminescence decay fitting, values of ρ = 2.1x 10^{-4}, 5.05 x 10^{-4} and 2.1 x 10^{-3} $N_{acceptors}$/ Å^3 can be calculated in polymers **(VII), (VIII)** and **(IX)** respectively.

Regarding the energy transfer process which quenches progressively the emission of the excited $Re(CO)_3(tmphen)^+$ (donor D) pendants with the increasing number of $Re(CO)_3(NO_2\text{-phen})^+$ (acceptor A) pendants in polymers **(VII)-(IX)**, the crucial point is that when passing from polymer **(V)** to **(VII)** the probability of a donor D to be in the vicinity of an acceptor A increases.

The efficiency of energy transfer between D and A in polymers **(VII)-(IX)** can be calculated according to a modification of eq. 25, i.e. eq. 28:

$$E_T^j = 1 - \frac{\phi_D^j}{\phi_{mb}^j} \tag{28}$$

Where E_T^j represent the energy transfer efficiency in polymer j [j = **(VII), (VIII)** and **(IX)**], ϕ_D^j represents emission quantum yields for polymers **(VII), (VIII)** and **(IX)** and ϕ_{mb}^j are the emission quantum yields of the molar blends 90%-**(V)** + 10%-**(VI)**, 75%-**(V)** + 25%-**(VI)** and 50%-**(V)** + 50%-**(VI)**, respectively. The experimental E_T^j values for polymers **(VII), (VIII)** and **(IX)** are 0.58, 0.85 and 0.94 respectively (Table 3).

Taking the E_T^j calculated above for polymers **(VII)-(IX)**, mean values for the energy transfer rate constant between $MLCT_{Re\rightarrow tmphen}$ and $MLCT_{Re\rightarrow NO2\text{-}phen}$ can be calculated in these polymers with the aid of eq. 29:

$$\overline{k_{ET}} = \frac{1}{\tau_D}\left(\frac{E_T^j}{1-E_T^j}\right) \quad (29)$$

The values of $\overline{k_{ET}}$ obtained for polymers **(VII), (VIII)** and **(IX)** are $2.7\text{x}10^5$, $1.1\text{x}10^6$ and $3.1\text{x}10^6$ s^{-1}, respectively. It should be noted that those $\overline{k_{ET}}$ values are only average contributions in the polymers. In fact, in evaluating k_{ET}, it has to be assumed that the quenching of the donor D is brought about by interactions with neighboring regions of the polymer containing the acceptor molecules A. The influence of these interactions depends on the distance between D and A. Then the quenching rate for the ith donor D_i is the sum over the distance dependent interactions with j quenching sites and $\overline{k_{ET}}$ represents a mean value over all possible $k_{ET}(D_i)$ (eq. 21).

Using E_T^j values for polymers **(VII)-(IX)** mean interchromophore distances can be calculated according to eq. 30:

$$\overline{R} = R_F\left(\frac{1}{E_T^j} - 1\right)^{1/6} \quad (30)$$

The values of $\overline{R}$ obtained for polymers **(VII), (VIII)** and **(IX)** are 10.1, 8.0 and 6.7 Å, respectively.

Flash photolysis experiments (λ_{exc} = 351 nm) on polymers **(V), (VII), (VIII)** and **(IX)** demonstrated that the $\Delta A_{t=0}$ values decrease from **(V)** to **(IX)** even when $\Delta A_{t=0}$ values are corrected for the decrease in n (i.e. from ~200 in polymer (V) to ~100 in polymer **(IX)**). For instance: $\Delta A_{t=0}$ **(V)**/ $\Delta A_{t=0}$ **(VII)** =1.7; $\Delta A_{t=0}$ **(V)**/ $\Delta A_{t=0}$ **(VIII)** =3.4; $\Delta A_{t=0}$ **(V)**/ $\Delta A_{t=0}$ **(IX)** = 5.2 while $n_{(V)}/n_{(VII)}$, $n_{(V)}/n_{(VIII)}$ and $n_{(V)}/n_{(IX)}$ are 1.1, 1.3 and 2 respectively. $\Delta A_{t=0}$ may be considered proportional to the photogenerated concentration of the MLCT excited states in flash fotolysis experiments on polymers **(V), (VII), (VIII)** and **(IX)**. This comparison suggests that the luminescence quenching occurs within the laser pulse lifetime (~25ns) when excited $Re(CO)_3(tmphen)^+$

pendants are situated at distances below R_F (~10.7 Å) from a $Re(CO)_3(NO_2$-phen$)^+$ acceptor and a manifestation of this "instant" quenching is the decrease of the $\Delta A_{t=0}$ values when comparing the polymers **(V)-(IX)** in a higher extent than $n_{(III)}/n_{(V)}$, $n_{(III)}/n_{(VI)}$ and $n_{(III)}/n_{(VII)}$. $Re(CO)_3$(tmphen$)^+$ excited pendants situated at distances > R_F from the acceptors $Re(CO)_3(NO_2$-phen$)^+$ will be quenched with rate constants $k_{ET}(D_i)$ and will be observed decaying at times longer than the laser lifetime. The overall decay of all $Re(CO)_3$(tmphen$)^+$ excited pendants situated at distances > R_F with all possible $k_{ET}(D_i)$ values will manifest in a luminescence decay according to eq. 22.

The transient spectra of the polymers **(V), (VII), (VIII)** and **(IX)** decayed biexponentially over a period of several microseconds. Even the decay of the transient of polymer **(V)** can not be fitted to a single exponential. However, the luminescence decay of polymer **(V)** after excitation with a N_2 laser (λ_{exc} = 337 nm) is monoexponential. Flash photolysis experiments were performed using an excimer laser (λ_{exc} = 351 nm) while flash fluorescence experiments (λ_{exc} = 337 nm) were performed using a N_2 laser. The energy/pulse from the excimer laser is ~15 times higher than that in flash fluorescence experiments. As a consequence, much higher concentrations of MLCT are produced in flash photolysis experiments than in flash fluorescent ones. We have previously observed marked differences (i.e. biexponenetial in contrast to monoexponential decays) when the MLCT excited state of the polymer **(II)** were generated in either high or low concentrations [23]. Such differences were related to MLCT annihilation processes in addition to the first order decay due to the presence of excited chromophores in close proximity in the polymer [21, 23]. The existence of Re(I) chromophores in diverse environments was shown by the intrinsic kinetics of the luminescence, the decay kinetics of the MLCT excited states observed by time resolved-absorption spectroscopy, and the quenching of the luminescence by various quenchers, with related mixed polymers **(III)** and **(IV)**. These experimental observations account for the presence of medium-destabilized charge-transfer excited states, in some chromophores within a strand of polymer [20]. Therefore, as the MLCT decay in polymers **(V), (VII)-(IX)** is much more complex in flash photolysis experiments than in flash fluorescent ones due to the higher concentrations of excited chromophores generated in the former experiments, the absorbance decay of the generated transients was fitted to a biexponential decay rather than that of eq. 22 used to fit luminescence profiles decays. As it can be observed from Table 3, the two lifetimes τ_{fast} and τ_{slow} decrease from polymers **(VII)** to **(IX)** in accordance with the increase in parameter a obtained from luminescence decay profiles. However, we are aware that the transient decay probably is not a truly biexponential and the two lifetimes may be corresponding to a random distribution of decay times which are approximately fitted to a sum of two exponentials and the two lifetimes may represent average contributions from excited chromophores in different environments [50].

Chapter 10

CONTRASTING INTRASTRAND PHOTOINDUCED PROCESSES IN MACROMOLECULES CONTAINING PENDANT $-Re(CO)_3$(1,10-PHENANTHROLINE)$^+$: ELECTRON VERSUS ENERGY TRANSFER IN POLYMERS (I) AND (X)

SOLUTION CHEMISTRY

The UV-vis spectrum of polymers **(I)** and **(X)** in CH_3CN exhibited noticeable differences in the 400-500 nm region, i.e., in a region where the absorption bands due to the Re to phen charge transfer optical transitions are positioned, figure 20. In order to make the spectrum of **(I)** comparable with the spectrum of **(X)** in the region of the Re(I) to phen charge transfer transitions, the extinction coefficients in figure 20 were calculated on the basis of the concentration of $-Re(CO)_3(phen)^+$ chromophores. On the basis of this normalization, the charge transfer absorption band of **(X)** appears 30-50 nm red-shifted with respect to the similar absorption band in the spectrum of **(I)**. Differences were also observed between the emission spectra of **(I)** and **(X)** when deaerated solutions in CH_3CN of either polymer (~4 x 10^{-5} M in Re(I) chromophore) were steady state irradiated at 350 nm. Since charge transfer transitions are sensitive to the environment, the distinct environments around the $-Re-(CO)_3(phen)^+$ chromophores, respectively created by the pendants py and $py-CH_3^+$ in **(I)** and **(X)**, must be the cause of the differences in the absorption spectrum. In accordance with this proposition, the observed differences in the emission spectra suggest that distinctive environments of the electronically excited Re(I) chromophores in **(I)** and **(X)** affect the decay of the emissive excited state.

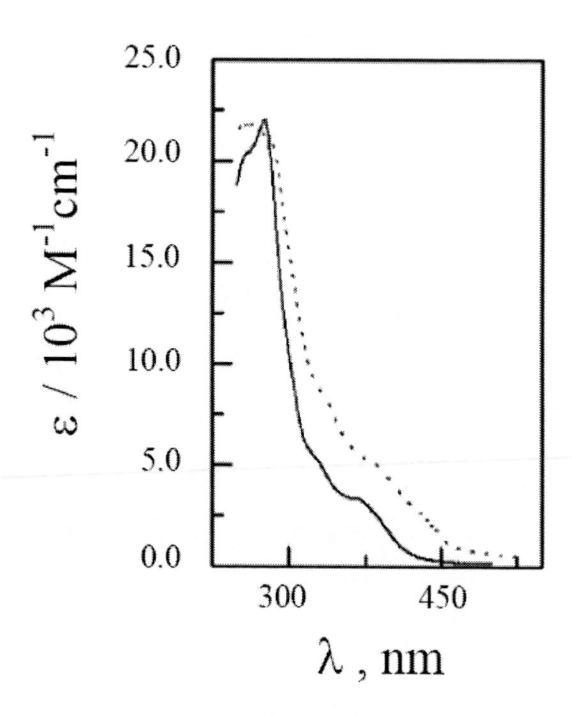

Figure 20. UV-vis spectra of polymers (I) (—) and (X) (---) in CH_3CN. The extinction coefficients were calculated as A/[-Re(CO)$_3$(phen)$^+$], where [-Re(CO)$_3$(phen)$^+$] is the concentration of Re(I) chromophore in the solutions of polymers (I) or (X).

Time-Resolved Spectroscopy

To investigate the time-resolved emission of **(X)**, 7.2×10^{-7} M deaerated solutions of the polymer (~1.4×10^{-4} M in Re(I) chromophore) in CH_3CN were flash irradiated at 350 nm. The oscillographic traces collected in experiments where the emission was monitored over a wide range of wavelengths, $550 \leq \lambda_{ob} \leq 610$ nm, were fitted to a bi-exponential decay, $A_1 \exp(-t/\tau) + A_2 \exp(-t/\tau')$. The lifetimes calculated for the decay of the emission at $\lambda_{ob} = 580$ nm were $\tau = 50 \pm 3$ ns and $\tau' = 1.45 \pm 0.08$ μs, and the relation of the pre-exponential factors was $A_1/A_2 = 0.5$, figure 21. A minor dependence of

τ' on λ_{ob} was also observed but was more pronounced in the decay of the transient absorption spectrum described below. In addition, the bi-exponential decay of the luminescence of **(X)** contrasts with the single exponential decay of the electronically excited chromophores in **(I)** (τ = 0.86±0.05 μs). Three well differentiated steps were observed in the decay of the transient absorption spectrum recorded 10 ns after a deaerated solution containing 7.2 x 10^{-7} M of **(X)** (~1.4 x 10^{-4} M in Re(I) chromophore) in CH_3CN was flash irradiated at 350 nm, figure 22. A broad absorption band with a maximum at 450 nm was observed in the transient spectrum recorded with delays (from the 10 ns flash irradiation) equal to or less than 2 μs. The transient absorption spectrum underwent a partial bi-exponential decay with a lifetime, τ = 48 ±2 ns, independent of the monitoring wavelength and τ' increasing from 1.12 ±0.06 μs (λ_{ob} = 410 nm) to 2.08 ± 0.06 (λ_{ob} = 600 nm). An average value, $<\tau'>$ = 1.7 ±0.3 μs, was calculated from the values collected between 410 and 600 nm. The lifetimes of the transients are nearly the same calculated for the time-resolved luminescence and must be related to the decay of excited states. Therefore, the lifetimes from the transient absorption spectra were assigned to the decay of the IL and the $MLCT_{Re\rightarrow phen}$ excited states, as was in the case of some monomeric Re(I) complexes [21,52,53]. By contrast to the photo behavior of the monomeric Re(I) complexes, an additional transient was observed after the decay of the excited states of **(X)**. The long-lived transient, poly-(LL), fulfills the spectroscopic and kinetic conditions of a displacement of charge through the strand of polymer, i.e., one that occurs during the decay of the MLCT and induces the creation of a Re(II) center at a distance from the $phen^{\bullet -}$ ligand radical. The absorption spectra recorded with delays longer than 2 μs showed a maximum at 500 nm, i.e., where a $phen^{\bullet -}$ ligand radical coordinated to Re(I) instead of Re(II) has its absorption maximum [21,52,53]. The kinetics of the optical density change, investigated between 370 and 520 nm, showed that the formation of the radical and the decay of the MLCT excited states are simultaneous processes. Indeed, the shift of the absorption maximum from 450 to 500 nm and the decay of the $MLCT_{Re\rightarrow phen}$ excited state, followed at 410 nm, have nearly the same lifetimes, ≈1.55 ±0.07 μs. The disappearance of the poly-(LL) transient spectrum, ascribed mainly to the $phen^{\bullet -}$ ligand radical, was followed by means of the optical density change, ΔA, at 500 nm. It was fitted to a single exponential, $\Delta A = \Delta A_0 \exp(-t/\tau'')$, with a lifetime τ'' = 8.0 μs. In agreement with the a separation of charge in the MLCT first and in a charge separated intermediate later, the flash irradiation induced a prompt bleach of the optical density at λ_{ob} = 380 nm, i.e., the region of the Re(I) to phen charge-transfer absorption band in **(X)**. The single exponential fitted to the recovery of 370 nm optical density has a lifetime, τ =7.7 ±0.7 μs, very close to the value reported above for the decay of the poly-(LL) intermediate with λ_{max} = 500 nm. In flash luminescence experiments carried out under similar experimental conditions, no luminescence between 300 and 600 nm could be observed in the time scale corresponding to the final decay of the transient spectrum. The absence of a long-lived luminescence with an ≈ 8.0 μs lifetime confirms that the transient spectrum must be assigned to a reaction intermediate instead of an excited state. Since the spectral changes in the longest time scale are in accordance with those of a species containing separated Re(II) and phen$^{\bullet-}$ radical moieties, it is possible to account for the decay kinetics by considering that the separated Re(II) and phen$^{\bullet-}$ moieties undergo charge annihilation processes within the strand of polymer. It was demonstrated that the formation of poly-(LL) was not related to a redox reaction of the excited states with the solvent by using solutions containing 7.2 x 10^{-7} M of **(X)** (~1.4 x 10^{-4} M in Re(I) chromophore) in a 50% (v/v) CH_3-OH/CH_3CN mixed solvent. Time-resolved spectral changes similar to those seen when the solvent was CH_3CN, Figure 22, were observed when deaerated solutions in the mixed solvent were flash photolyzed at 350 nm.

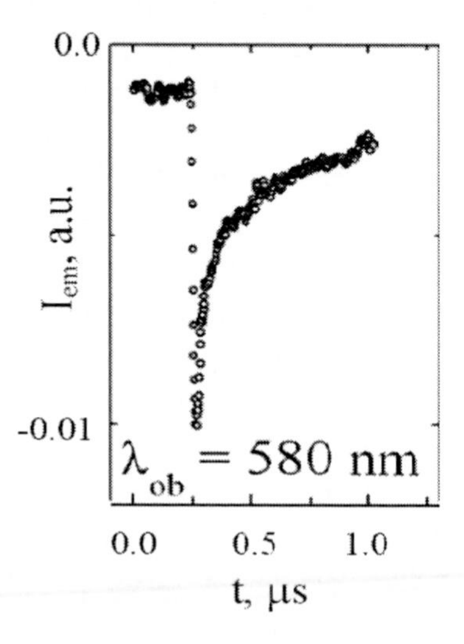

Figure 21. A typical trace for the decay of the 580 nm luminescence recorded when a deaerated solution of (X) (~1.4 x 10^{-4}M in Re(I) chromophore) in CH_3CN was flash irradiated at 350 nm.

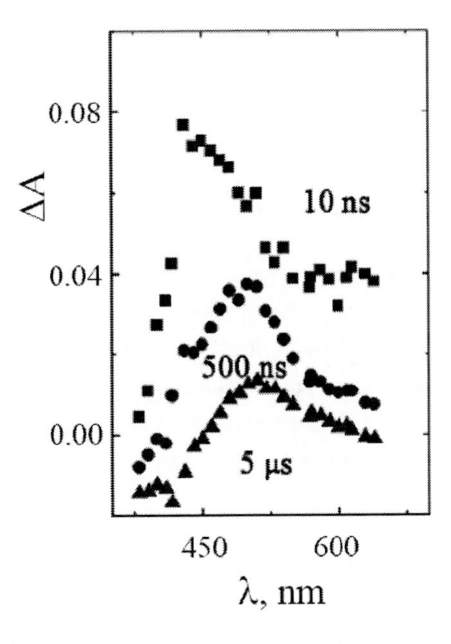

Figure 22. Transient spectra recorded in the 350 nm flash irradiation of deaerated solutions 7.2 x 10^{-7}M in (X) (~1.4 x 10^{-4} M in Re(I) chromophore) in CH_3CN. The delays from the laser irradiation are indicated in the figure.

QUENCHING REACTIONS

The experiments communicated above revealed some distinct photochemical properties of $pyRe(CO)_3(phen)^+$ and the polyelectrolytes **(I)** and **(X)**. To investigate how the different environments of the chromophores, i.e., the presence of pyridinium versus pyridine groups, affected the reactivity of the $MLCT_{Re\rightarrow phen}$, the excited state reactions were investigated with solutions of **(I)** and **(X)**, ~1 x 10^{-6} M in polyelectrolyte, containing counterions more reactive than triflate. In order to have adequate solubilities of the polyelectrolytes and their counterions, the solutions for the photochemical work had to be prepared in a 50% (v/v) CH_3OH/CH_3CN mixed solvent. Solutions where the concentration of I^- or SO_3^{2-} was increased from 1.0 x 10^{-4} to 1.0 x 10^{-2} M showed a progressive quenching of the $-Re(CO)_3(phen)^+$ luminescence in **(I)** or **(X)**. By contrast, very little quenching of the $pyRe(CO)_3(phen)^+$ luminescence was observed with solutions containing concentrations of the chromophore and the counterions similar to those of **(I)** and **(X)**. In flash fluorescence experiments with **(I)** and **(X)**, the lifetimes for the decay of the luminescence decreased with I^- concentration. To find the reason for the quenching of the -$Re(CO)_3(phen)^+$ luminescence, transient absorption spectra were recorded with various delays after the 350 nm flash irradiation of solutions having 1.2 x 10^{-2} or 1.3 x 10^{-3} M I^-, figure 23. While the prompt spectrum observed after the flash irradiation of the solution with 1.3 x 10^{-3} M I^- exhibited some of the features of the electronically excited chromophores, figure 23b, the spectra recorded with longer delays or with a solution containing I^- in higher concentrations agreed with the literature spectrum of I_2^-, figure 23a. Concentration of the I_2^- product per flash was calculated from the ΔA change at 400 nm, and the values of the absorbance change were recorded on a time scale where the decay of I_2^- was insignificant, e.g., delays $1.0 \leq t \leq 10$ μs after the flash irradiation. In accordance with a redox reaction of I^- with the electronically excited Re(I) pendant, concentrations of flash-generated I_2^- increased with I^- concentration, figure 24. Oscillographic traces, λ_{ob} = 400 nm, showing the decay of I_2^- were fitted to linear inverse plots, ΔA^{-1} versus t, inset to figure 24. It was concluded that the decay of I_2^- is kinetically second order under the experimental conditions used for these experiments. The rate constant, $2k$ = 8.6 x 10^9 M^{-1} s^{-1} at an ionic strength $I \approx 1$ M at the strand of polyelectrolyte, was calculated from the ratio of the rate constant to the extinction coefficient, $2k/\varepsilon$ = 6.2 x 10^5 M^{-1} s^{-1} at λ_{ob} = 400 nm, and the literature value of the I_2^- extinction coefficient, ε = 1.4 x 10^4 M^{-1} cm^{-1} at such a wavelength [54]. It is possible to associate the decay of I_2^- with the reoxidation of the $phen^{\bullet-}$ ligand-radicals that remain coordinated to Re(I) after the quenching of the $MLCT_{Re\rightarrow phen}$ excited state.

Other excited state reductants, i.e., 10^{-3} M SO_3^{2-}, 0.1 M TEOA, or 0.1 M TEA, react with the electronically excited chromophores. The transient absorption spectra, recorded when flash photolyzed solutions of **(I)** or **(X)** contained one of these quenchers, have the spectral features previously attributed to a Re(I)-coordinated $phen^{\bullet-}$ radical [21,52,53]. Therefore, we concluded that a reductive quenching of the $MLCT_{Re\rightarrow phen}$ excited state in **(I)** or **(X)** led to the spectroscopic changes. The concentrations of I^- and SO_3^{2-} in their redox reactions with the electronically excited Re(I) chromophores of **(I)** and **(X)** were ca. 2 orders of magnitude smaller than those required for the quenching of the electronically excited $pyRe(CO)_3(phen)^+$. A reason for this larger efficiency is a more favorable electrostatic interaction between the anionic quenchers and the cationic polyelectrolytes. A complete kinetic study of the reactions of the excited polymer with oxidants was hindered by limitations that optical conditions and the solubilities of the reactants imposed in these experiments. However, the oxidation of poly-(LL) by tetracyanoethylene, TCE, was observed under a limited range of TCE concentrations. Solutions of the triflate salt of **(X)** containing 1 x 10^{-6} M in polyelectrolyte and 1.0 x 10^{-4} M to 1.0 x 10^{-3} M TCE were flash irradiated at 350 nm. The disappearance of poly-(LL) with a lifetime $\tau \sim 2$ μs was observed at λ_{ob} = 500 nm.

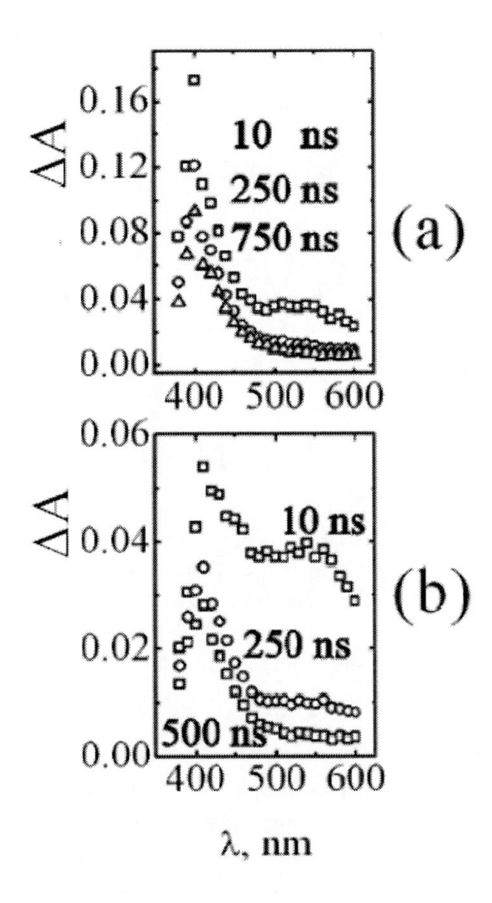

Figure 23. Transient spectra observed in the reaction of I^- (1.2 x 10^{-2} M top and 1.3 x 10^{-3} M bottom) with electronically excited polymer (X). The transients were recorded in the 350 nm flash irradiation of 7.2 x 10^{-7} M (X) (~1.4 x 10^{-4} M in Re(I) chromophore) in deaerated CH_3CN/CH_3OH (50% v/v). The delays after the laser irradiation are indicated by the side of the spectra.

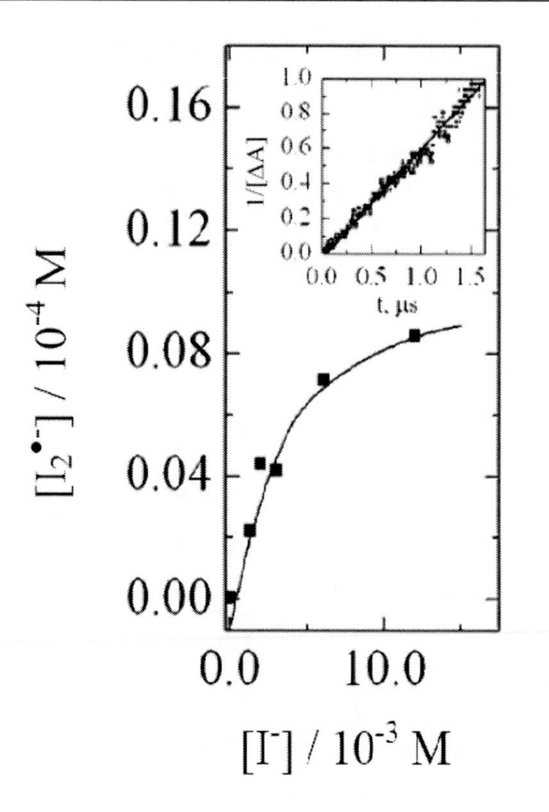

Figure 24. Concentrations of I_2^- from the reaction of I^- with electronically excited (X) are shown as a function of the I^- concentration. The 350 nm flash photolyzed solution also contained 7.2 x 10^{-4} M in Re(I) chromophore) in deaerated CH_3CN/CH_3OH (50% v/v). The I_2^- concentration were calculated from the ΔA observed 250 ns after the laser irradiation at 400 nm. The inset shows an inverse plot of the ΔA at 400 nm for the back electron-transfer reaction of I_2^-.

The experimental observations in this work and in recent literature reports show that Re(I) sensitizers imbedded in polymeric structures provide routes for new photoinduced reactions that could be more efficient than those available to the monomeric moieties [21, 53]. Some of the routes appear to be controllable by the environs that surround the chromophore, e.g., the Re(I) chromophores in **(I)** and **(X)**. Since there are sharp differences between the photobehavior of **(X)** and the literature reported photobehavior of **(I)** [21], the 350 nm photochemistry of **(X)** will be discussed first and the differences between the photobehavior of **(I)** and **(X)** will be rationalized at the end of the section.

PHOTOPHYSICS OF THE ELECTRONICALLY EXCITED (X)

The results of time-resolved and steady state experiments can be properly modeled on the basis of eqs 31-41 which are an abbreviated form of those shown in the scheme of figure 25.

$$\textbf{(X)} + h\nu \rightarrow \text{poly-(IL)};\ \varphi_1 \quad (31)$$

$$\textbf{(X)} + h\nu \rightarrow \text{poly-(MLCT)};\ \varphi_2 \quad (32)$$

$$\text{poly-(IL)} \rightarrow \text{poly-(MLCT)};\ k_1 \quad (33)$$

$$\text{poly-(IL)} \rightarrow \textbf{(X)};\ k_{2,nr} \quad (34)$$

$$\text{poly-(IL)} \rightarrow \textbf{(X)};\ k_{2,r} \quad (35)$$

$$\text{poly-(MLCT)} \rightarrow \textbf{(X)};\ k_{3,nr} \quad (36)$$

$$\text{poly-(MLCT)} \rightarrow \textbf{(X)} + h\nu';\ k_{3,r} \quad (37)$$

$$\text{poly-(MLCT)} + Q \rightarrow \text{products};\ k_{R1} \quad (38)$$

$$\text{poly-(MLCT)} \rightarrow \text{poly-(LL)};\ k_4 \quad (39)$$

$$\text{poly-(LL)} \rightarrow \textbf{(X)};\ k_5 \quad (40)$$

$$\text{poly-(LL)} + Q \rightarrow \text{products};\ k_{R2} \quad (41)$$

The species poly-(IL) and poly-(MLCT) represent strands of **(X)** populated with different excited states. They are generated with quantum yields φ_1 and φ_2 by the absorption of light, eqs 31 and 32, and decay via radiative, eqs 35 and 37, and radiationless, eqs 34 and 36, processes. On the basis of the experimental observations, the transformation of the poly-(MLCT) into the poly-(LL) (eq. 39) intermediate has to be represented as a process that is kinetically first order in the excited chromophores. If pairs of excited chromophores are converted to poly-(LL) as radicals are converted to products in a geminate pair, the population of excited states in poly- (MLCT) must decay exponentially in time. It is also possible that MLCT excited states in the strand are placed in different environments and undergo bimolecular transformations to poly-LL with different rates [20]. A convolution function describing the overall decay of the excited state concentration will approximately have the shape of an exponential. Although the dependence of the lifetime on λ_{ob} is consistent with an assortment of decay rates, the possibility of pairs of excited states decaying exponentially in time cannot be ruled out.

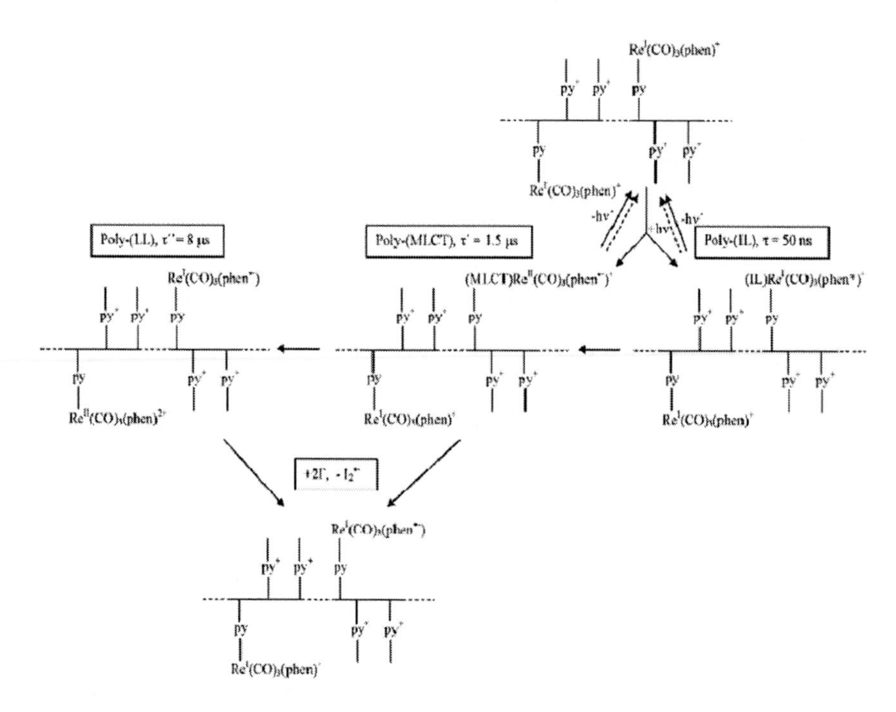

Figure 25. Scheme showing formation of poly-(IL) and poly-(MLCT) after optical excitation. Intra-strand electron transfer processes lead from poly-(MLCT) to poly-(LL). Reactions of poly-(LL) and poly-(MLCT) with I^- yields $I_2^{\bullet -}$.

In addition, poly-(LL) and poly-(MLCT) undergo electron transfer reactions with a reactant Q, eqs 38 and 41. Integration of the rate equations derived for the mechanism in eqs 31-41 with [poly-(IL)]$_0$ and [poly-(MLCT)]$_0$ being the flash-generated concentrations of poly-(IL) and poly- (MLCT) gives the following relationships for the emission intensity, eq 42, and the quotient of pre-exponential factors, eq 43, when the quencher, Q, concentration is [Q] = 0.

$$I_{em,t} = I_{em,0}\left[\frac{\rho}{1+\rho}\exp(-t/\tau) + \frac{1}{1+\rho}\exp(-t/\tau')\right] \tag{42}$$

$$\frac{A_1}{A_2} = \frac{\dfrac{k_{2,r}}{k_{3,r}} + \dfrac{k_1\tau\tau'}{\tau-\tau'}}{\dfrac{[poly-MLCT]_0}{[poly-IL]_0} + \dfrac{k_1\tau\tau'}{\tau-\tau'}} = \rho \tag{43}$$

With $\tau = (k_1 + k_{2,nr} + k_{2,r})^{-1}$ and $\tau' = (k_4 + k_{3,nr} + k_{3,r})^{-1}$

Insertion of the experimental result $A_1/A_2 = 0.5$ in eqs 42 and 43 gives the expression for the emission intensity, eq 44.

$$I_{em,t} = I_{em,0}\left[0.33\exp(-t/\tau) + 0.66\exp(-t/\tau')\right] \tag{44}$$

In accordance with eq 42, the lifetimes calculated from the time-resolved luminescence experiments must be $\tau = (k_1 + k_{2,nr} + k_{2,r})^{-1} \sim$ 50 ns and $\tau' = (k_4 + k_{3,nr} + k_{3,r})^{-1} \sim$ 1.45 μs. With exclusion of the formation of the LL intermediate, the photophysical processes of the IL and $MLCT_{Re\rightarrow phen}$ excited states in the scheme resemble those reported in the literature for polymer **(I)** [21, 26]. On the basis of the luminescence results, the IL and $MLCT_{Re\rightarrow phen}$ excited states are simultaneously generated with the absorption of 350 nm light. The short-lived decay of the luminescence, τ = 50 ns, is spectroscopically and kinetically consistent with the decay of an IL excited state placed above the $MLCT_{Re\rightarrow phen}$. Previous works have shown that IL excited states undergoes competitive decays to the lower lying $MLCT_{Re\rightarrow phen}$ and to the ground state [53, 55, 56]. The subsequent process with a lifetime τ'= 1.45 μs is kinetically and spectroscopically consistent with the relaxation of the $MLCT_{Re\rightarrow phen}$. Indeed, the 1.45 μs lifetime is nearly the same lifetime of the $MLCT_{Re\rightarrow acceptor}$ in a related monomeric complex [53]. The subsequent process has a lifetime, τ''= 8.0 μs, and it is too slow to be ascribed to the room temperature radiationless relaxation of an excited state in a triscarbonyl Re(I) complex. A charge-separated species with a Re(II) chromophore and a reduced phen, i.e., a $phen^{\bullet -}$ radical separated of the Re(II) metal center, provides a better rationale for the transient absorbing with a $\lambda_{max} \approx 500$ nm and a lifetime, $\tau'' = 8.0\ \mu s$. Such a species, see Scheme in figure 25, must be produced by a remote displacement of charge within the strand of polymer, i.e., a reaction that competes with the radiative and radiationless relaxations of the $MLCT_{Re \rightarrow phen}$.

*Contrasting Photoprocesses of **(I)** and **(X)**.* The most evident difference between the photophysics of **(X)** and **(I)** is that a remote charge transfer operates in the former while excited state annihilations processes, due to energy migration in the strand of polymer, are favored in the latter. It is expected that the different intrastrand processes are a consequence of the different environmental conditions respectively created around the $MLCT_{Re \rightarrow phen}$ by the pyridine and pyridinium pendants. Migration of energy in **(I)** can be expected to be a collisional process, i.e., a Dexter's exchange mechanism [32], requiring that two Re(I) pendants come to a close proximity. Intrastrand encounters of this nature must be hindered in **(X)** because of electrostatic interactions between the cationic Re(I) and pyridinium pendants. However, pathways for a remote electron transfer could be available in **(X)**. The formation of poly-LL through these paths can be fast enough to compete with the relaxation of the $MLCT_{Re \rightarrow phen}$ excited state. Absence of delayed luminescence with an 8.0 μs lifetime shows that back-electron transfers in poly-LL to restore the $MLCT_{Re \rightarrow phen}$ or generate other luminescent excited states are inefficient processes. Therefore, the lifetime of poly-LL must be controlled by the radiationless decay to the ground state. Since the donor-acceptor distances for the charge migration in the decay of poly-LL to the ground state or to the $MLCT_{Re \rightarrow phen}$ are nearly the same, it cannot be the reason for the difference between the reaction rate constants of these two processes [57]. Quenching experiments indicate that Re(I) chromophores in **(I)** and **(X)** must be inside the positive electrostatic field of the strand. Indeed, the anionic reactants, I^- or SO_3^{2-}, are more efficient electron transfer quenchers of the $MLCT_{Re \rightarrow phen}$ luminescence in **(I)** and **(X)** than they are for the excited state of $pyRe(CO)_3(phen)^+$. Relative to the monomeric complex, the enhanced efficiency of the quenchers must be due to stronger electrostatic interactions between the reactants, with the anions possibly located in the double layer of the polyelectrolyte. When the electron transfer quenching was followed by means of the concentration of the I_2^- product, the charge-separated species with a longer lifetime, $\tau'' = 8\ \mu s$, than the $MLCT_{Re \rightarrow phen}$ excited state, $\tau' = 1.5\ \mu s$, appears to be reduced by I^-, eq 38. Moreover, the mechanism, eqs 31-41, predicts that the I_2^- concentration will approach the limiting value [poly-

MLCT]$_0$ + [poly-IL]$_0$, i.e., the trend seen in figure 24 for the dependence of the I_2^-concentration on I^-.

Chapter 11

CONCLUSIONS

Inorganic polymers with general formula $\{[(vpy)_2\text{-}vpyRe(CO)_3L]CF_3SO_3\}_{n\sim200}$ may be easily synthesized from the application of ligand substitution reactions of the Re(I) carbonyl–diimine *fac*-$Re(L)(CO)_3(\alpha\text{-diimine})^{0/+}$ complexes to the Poly-4-(vinylpyridine) backbone. These polymers can be assigned to a formal structure where pendant -$Re(CO)_3L^+$ groups are randomly distributed along the strand of polymer with an average of two 4-vinylpyridine groups for each one coordinated to a Re(I) chromophore. The outcome of this spatial distribution of Re(I) pendants is that distances as low as 8Å may be encountered between vicinal metallic centers. This close vicinity of Re(I) pendants exerts a profound influence on the photophysical and photochemical properties of these polymers.

The morphology of polymers was studied by TEM. The polymer films were obtained by room temperature solvent evaporation of its acetonitrile or dichloromethane solutions. The Re(I) complexes in the polymer aggregate and form isolated nanodomains that are dispersed in the polymer matrix film. By using different solvents multiple morphologies of aggregates from these Re(I) polymers were obtained ranging from spherical micellar-like objects to branched tubular structures intertwined in a net and large compound vesicles. Similar results were obtained from dynamic and static light scattering measurements indicating that the nanoaggregates also exist in solution. Also, the specific environments that the strands and aggregates of strands create for the Re(I) chromophores affect their photochemical properties (e.g.,the lifetime of the MLCT excited states). Moreover, marked differences were found between the photochemical and photophysical properties of the polymers and those of the related monomeric complexes, $pyRe(CO)_3L^+$. For instance, in polymers $\{[(vpy)_2\text{-}vpyRe(CO)_3L]CF_3SO_3\}_{n\sim200}$ the annihilation of two ^{3}MLCT excited states, forms chromophores in the ground state and intraligand, IL, electronic state. The latter being an excited state with a higher energy than the ^{3}MLCT excited state. In contrast to the ^{3}MLCT excited states, the ^{3}IL excited states oxidize organic solvents (e.g., CH_3OH). These processes were not observed with the related monomeric complexes $pyRe(CO)_3L^+$. The main cause of these differences is the photogeneration of MLCT excited sates in concentrations that are much larger when $\text{-}Re(CO)_3\ L^+$ chromophores are bound to $(vpy)_{n\sim600}$. This is the photophysical result of Re(I) chromophores being crowded in strands of a polymer instead of being homogeneously distributed through solutions of a $pyRe(CO)_3L^+$ complex. Irradiations at 350 nm induce intrastrand charge separation in the peralkylated polymer, $[(vpy\text{-}CH_3^+)_2\text{-}vpyRe(CO)_3(phen)]_{n\sim200}$, a process that stands in contrast with the energy migration observed with $[(vpy)_2\text{-}vpyRe(CO)_3(phen)]_{n\sim200}$. Electronically excited $\text{-}vpyRe(CO)_3(phen)^+$ chromophores and charge separated intermediates react with neutral species, e.g., TEOA, and anionic electron donors, e.g., SO_3^{2-} and I^-. The anionic electron donors react more efficiently with the MLCT excited state of these polyelectrolytes than with the MLCT excited state of $pyRe(CO)_3(phen)^+$. The different photoreactions of $[(vpy)_2\text{-}vpyRe(CO)_3(phen)]_{n\sim200}$ and $[(vpy\text{-}CH_3^+)_2\text{-}vpyRe(CO)_3(phen)]_{n\sim200}$ suggest that a more diverse photobehavior should be attained when pyridine and or pyridinium pendants are replaced by ionic electron donor/acceptors. Diverse molecular architectures can be devised in order to induce photoreactions that are atypical in the photochemistry of the isolated pendants. Among the various photoinduced processes of these newer macromolecules, there are sequential multielectron processes of practical and theoretical interest.

Energy transfer between MLCT excited states inside mixed polymers like $\{[(vpy)_2vpyRe(CO)_3(tmphen)^+]\}_n\{[(vpy)_2vpyRe(CO)_3(NO_2\text{-}phen)^+]\}_m$ was evidenced by steady state and time-resolved spectroscopy. The substitution of pendants $Re(CO)_3(tmphen)^+$ by pendants $Re(CO)_3(NO_2\text{-}phen)^+$ in the poly-4-vynilpyridine backbone produces a decrease in ϕ_{em} that is in proportion to the number of $Re(CO)_3(NO_2\text{-}phen)^+$ pendants relative to that of $Re(CO)_3(tmphen)^+$ pendants due to an energy transfer process that involves the excited states $MLCT_{Re\rightarrow\ tmphen}$ and $MLCT_{Re\rightarrow\ NO2\text{-}phen.}$ Current Förster resonance energy transfer theory was successfully applied to energy transfer processes in these polymers.

ACKNOWLEDGMENTS

The authors acknowledge support from ANPCyT Grant No. PICT 26195, CONICET-PIP 6301/05, Universidad Nacional de La Plata, and CICPBA. E.W. is a member of CONICET and M.R.F. is a member of CICPBA.

References

[1] Ruiz, G.; Wolcan, E.; Féliz, M.R. *J. Photochem Photobio. A: Chem* 1996, *101*, 119-125.

[2] Stufkens, D. J.; Vlček, A. Jr. *Coord Chem Rev* 1998, *177*, 127–179.

[3] Fox, M. A. and Chanon, M. *Photoinduced Electron Transfer;* Elsevier, Amsterdam, 1988.

[4] Balzani, V.; Bolleta, F.; Gandolfi, M. T.; Maestri, M. *Top. Curr. Chem.*,1978, *75*, 1.

[5] Grätzel, M. *Energy Resources Through Photochemistry and Catalysis,* Academic Press, New York, 1983.

[6] Kalyanasundaram, K. *Coord Chem Rev*, 1982, *46*, 159.

[7] Kalyanasundaram, K.; Grätzel, M. *Photosensitization and Photocatalysis Using Inorganic and Organometallic Compounds,* Kluwer Academic Publishers, Dordrecht, 1993.

[8] Yam, V.W.-W.; Wong, K.M.-C.; Lee, V.W.-M.; Lo, K.K.-W.; Cheung, K.-K. *Organometallics* 1995, *14*, 4034.

[9] Sacksteder, L.; Lee, M.; Demas, J.N.; DeGraff, B.A. J. *Am Chem Soc* 1993, *115*, 8230.

[10] Yoon, D.I.; Berg-Brennan, C.A.; Lu, H.; Hupp, J.T. *Inorg Chem* 1992, *31*, 3192.

[11] Berg-Brennan, C.A.; Yoon, D.I.; Slone, R.V. ; Kazala, A.P.; Hupp, J.T. *Inorg Chem* 1996, *35*, 2032.

[12] Calabrese, J.C.; Tam, W. *Chem Phys Lett* 1987, *133*, 244.

[13] Ehler, T.T.; Malmberg, N.; Carron, K.; Sullivan, B.P.; Noe, L.J. *J. Phys Chem B* 1997, *101*, 3174.

[14] Yam, V.W.-W.; Lau, V.C.Y.; Cheung, K.-K. *J. Chem Soc Chem Commun* 1995, 259.

[15] Huynh, M. H. V.; Dattelbaum, D. M. ;Meyer, T.J. *Coord Chem Rev* 2005, *249*, 457–483.

[16] Connors, P. J. Jr.; Tzalis, D.; Dunnick, A. L.; Tor, Y. *Inorg Chem*, 1998, *37*, 1121.

[17] Turro, N. J. *Modern Molecular Photochemistry*; University Science Books: Mill Valley, CA, 1991; Chapter 9, p 296.

[18] Zingales, F.; Satorelli, L.; Trovati, A. *Inorg Chem*, 1967, *6*, 1246

[19] Wolcan, E. ; Alessandrini, J. L.; Féliz, M.R. J. Phys. Chem. B, 2005, *109*, 22890-22898.

[20] Wolcan, E. ; Féliz, M.R.; Alessandrini, J. L.; Ferraudi, G. *Inorg. Chem.* 2006, *45*, 6666-6677

[21] Wolcan, E.; Ferraudi, G. *J. Phys Chem A* 2000, *104*, 9281-9286.

[22] Ranby, R.; Rabek, J. F. *Photodegradation, Photooxidation and Photostabilization of Polymers. Principles and Applications*; Wiley-Interscience: New York, 1975; Chapter 7.

[23] Wolcan, E. ; Féliz, M.R. *Photochem. Photobiol. Sci.*, 2003, *2*, 412-417.

[24] Hou, S. ; Chan, W. K. *Macromol. Rapid Commun.* 1999, *20*, 440–443.

[25] Nakano, T.; Okamoto, Y. *Chem. Rev.* 2001, *101*, 4013.

[26] Wolcan, E. ; Ferraudi, G.; Féliz, M.R.; Gómez, R.V.; Mickelsons, L. Supramol. Chem. 2003, *15*, 143-148.

[27] Scatchard, G. *Ann. N.Y. Acad. Sci.* 1949, *51*, 660.

[28] McGhee, J. D.; von Hippel, P. H. *J. Mol. Biol.* 1974, *86*, 469.

[29] Agnew, H. *J. Polym. Sci.* 1976, 14, 2819.

[30] Kirsh, Y. E.; Kovner, V. Ya.; Kokorin, A. I.; Zamaraev, K. I.; Cherniak, V. Ya.; Kabanov, V. A. *Eur. Polym. J.* 1974, 10, 671.

[31] Ruiz, G.; Rodriguez-Nieto, F.; Wolcan, E.; Féliz, M. R. *J. Photochem. Photobiol. A: Chem.* 1997, 107, 47.

[32] Wilkinson, F. in *Photoinduced Electron Transfer*, Marye Anne Fox and Michel Chanon (Eds.), Elsevier, Oxford, 1988, p.208.

[33] Draxler, S. ; Lippitsch, M. E. *Anal. Chem.* 1996, *68*, 753.

[34] Wallace, L.; Rillema, D. P. *Inorg. Chem.* 1993, *32*, 3836.

[35] Busby, M.; Gabrielsson, A.; Matousek, P.; Towrie, M.; Di Bilio, A. J.; Gray, H. B.; Vlček, A. Jr. *Inorg. Chem.* 2004, *43*, 4994.

[36] Scholes, G. D. *Annu. Rev. Phys. Chem.* 2003, *54*, 57.

[37] Rolinski, O. J.; Mathivanan, C.; Macnaught, G.; Birch, D. J. S. *Biosensors Bioelectron.* 2004, *20*, 424.

[38] Selvin, P. R. *Nat. Struct. Biol.* 2000, *7*, 730.

[39] Salthammer, T.; Dreeskamp, H.; Birch, D. J. S.; Imhof, R. E. *J. Photochem. Photobiol. A: Chem.* 1990, *55*, 53.

[40] Birch, D. J. S.; Rolinski, O. J.; Hatrick, D. ReV. Sci. Instrum. 1996, 67, 2732.

[41] Birch, D. J. S.; Holmes, A. S.; Darbyshire, M. *Meas. Sci. Technol.* 1995, *6*, 243.

[42] Rhee, M. J.; Sudnick, D. R.; Arkle, V. K.; Horrocks, W. D., Jr *Biochemistry* 1981, *20*, 3328.

[43] Snyder, A. P.; Sudnick, D. R.; Arkle, V. K.; Horrocks, W. D., Jr *Biochemistry* 1981, *20*, 3334.

[44] Anni, M.; Manna, L.; Cingolani, R.; Valerini, D.; Cretí, A.; Lomascolo, M. *App. Phys. Lett.* 2004, *85*, 4169.

[45] Selvin, P. R. *Annu. Rev. Biophys. Biomol. Struct.* 2002, *31*, 275.

[46] Hemmila¨, I.; Laitala, V. *J. Fluoresc.* 2005, *15*, 529.

[47] Xiao, M.; Selvin, P. R. *Rev. Sci. Instrum.* 1999, *70*, 3877.

[48] Selvin, P. R. *IEEE J. Selected Top. Quantum Electron.* 1996, *2*, 1077.

[49] Selvin, P. R.; Tariq, M.; Rana, T. M.; Hearst, J. E. *J. Am. Chem. Soc.* 1994, *116*, 6029.

[50] Bracco , L. L. B.; Juliarena, M. P. ; Ruiz, G. T.; Féliz, M.R.; Ferraudi; G. J.; Wolcan, E. *J. Phys. Chem. B* 2008, *112*, 11506-11516.

[51] Huynh, M. H. V.; Dattelbaum, D. M.; Meyer, T. J. *Coord. Chem. Rev.* 2005, *249*, 457.

[52] a) Jones,W.E., Jr.; Hermans, L.; Jiang, B. In Multimetallic and Macromolecular Inorganic Photochemistry; Ramamurthy, V., Schanze, K.S., Eds.; Marcel Dekker Inc.; New York, 1999; Chapter 1, Vol. 4. b) Ogawa, M. Y. In Multimetallic and Macromolecular Inorganic Photochemistry; Ramamurthy, V., Schanze, K. S., Eds.; Marcel Dekker Inc. New ork, 1999; Chapter 3, Vol 4. c) Worl, L. A.; Jones, W. Jr.; Trouse, G. F.; Younathan, J.D.; Danielson, E.; Maxwell, K. A.; Sykora, M.; Meyer, T.J. Inorg.Chem. 1999, 38, 2705. d) Petersen, J.D. In Supramolecular Photochemistry; Balzani, V., Ed.; Reideel: Dordrecht, 1987; p 135. e) Balzani, V.; Scandola, F. In Supramolecular Photochemistry; Ellis Harwood; Chichester, England, 1991; p 355. f) Kaneko, M.; Tsuchida, E. J.Polym.Sci. 1981, 16, 397.

[53] Guerrero, J.; Piro, O.E.; Wolcan, E.; Feliz, M.R.; Ferraudi, G.; Moya, S.A. Organometallics 2001, 20, 2842.

[54] Hugh, G.L. Optical Spectra of Nonmetallic Inorganic Transient Species in Aqueus Solutions; nat. stand. Ref. Data ser.; NSRDS-NBS 69, 1981 and references therein.

[55] Feliz, M.; Ferraudi, G. J. Phys. Chem. 1992, 96, 3059.

[56] Feliz, M.; Ferraudi, G.; Altmiller, H. J. Phys. Chem. 1992, 96, 257.

[57] Moser, C. C.; Keske, J. M.; Warncke, K.; Farid, R. S.; Dutton, P. L. Nature 1992, 355, 79.

INDEX

F

G

H

I

K

L

M

N

O

P

Q

R

S

T

V

W